A SMART SUSTAINABLE PIPELINE DESIGN USING FREE ISRO'S BHUVAN SATELLITE DATA

A Remote Sensing GIS Approach

A SMART SUSTAINABLE PIPELINE DESIGN USING FREE ISRO'S BHUVAN SATELLITE DATA

A Remote Sensing GIS Approach

Dr. Karthik Nagarajan
Dr. S. K. Ukarande

NAMYA PRESS.com

Published by

Namya Press

USA: 10685-B Hazlehurst Dr. # 31013 Houston, Texas - 77043
Email: namyapress@gmail.com
Website: https://namyapress.com

Edition	:	First Published in 2022 (Namya Press)
Title	:	A Smart Sustainable Pipeline Design using Free ISRO's Bhuvan Satellite Data: A Remote Sensing GIS Approach
Editors	:	Dr. Karthik Nagarajan & Dr. S. K. Ukarande
ISBN	:	978-1-956861-19-8
Copyright	:	© Dr. Karthik Nagarajan & Dr. S. K. Ukarande 2022 All rights reserved

॥ माता पिता गुरु दैवम् ॥

mātā pitā guru daivam

Dedicated

to my

Parents

Appa

Mr. M. S. Nagarajan

Amma

Ms. Kalyani Nagarajan

Family

Teachers

Friends

Students

And

Well Wishers

Preface

It's with great happiness that I would like to acknowledge a lot of people that who have helped me extremely through the entire difficult, challenging, but rewarding and interesting path towards some sort of edited Book without having their help and support, none of this work could have been possible.

Dr. Karthik Nagarajan is highly indebted to Dr. Vasudevan Pillai, Secretary, and CEO of Mahatama Education Society (MES), Dr. Daphne Pillai, Joint Secretary, and Chairperson, Management Board of MES, Dr. Priam Pillai Chief Operating Officer (MES), Mr. Franav Pillai Director and Trustee at MES for their kind support and encouragement.

I would like to express my sincere gratitude to my supervisor Dr. Suresh. K. Ukarande (Principal - K.J. Somaiya Institute of Engineering and Information Technology, Sion, Mumbai) for the continuous support of my Ph.D. study and research. His patience, motivation, enthusiasm, immense knowledge, and guidance helped me throughout the research and writing of this thesis.

In this study, the Indian Institute of Remote Sensing, Indian Space Research Organisation (I.I.R.S., I.S.R.O.), Dehradun, India, supported by giving a platform for understanding remote sensing and G.I.S. and providing Spatial Satellite data of the study area through its web portal Bhuvan. City and Industrial Development Corporation (C.I.D.C.O.) supported providing nonspatial data of the study area, such as Land Use Land Cover (L.U.L.C.) maps and population details. In contrast, the toposheet of the study area was obtained from the Survey of India (S.O.I.), Dehradun.

I also thank Dr. Chelpa Lingam, Dr. Shrikant. B. Charhate, Dr. Madhumita Chatterjee, and Dr. T.J Mathew for always motivating and supporting me throughout my Ph.D. I am grateful to Dr. Jagdish.

W. Bakal (Principal – Pillai HOC College of Engineering and Technology, HOC Colony, Rasayni) for inspiring me.

Last but not least, I would like to thank my brother Mr.Balaji Nagarajan for always standing beside me and my family, friends, and all faculty of PHCET for supporting me spiritually and constantly encouraging me throughout the research work.

Contents

CHAPTER - 01

Introduction

1.1 GENERAL

India is the seventh-largest country by geographical area, second-most populous country, and the most populous liberal democracy globally. It is estimated that half of the Indian population would live in urban areas by the year 2050. The population residing in urban areas of the world (World Urbanization Prospects, United Nations,2014) would reach 70 % in 2050. A study (Goldman Sachs Economic Research,2011) estimated that 460 and 700 million people would move to Indian urban cities by 2020 and 2050, respectively. Timeline of Water Distribution System Modeling is shown in Table 1.1

Table 1.1: Timeline of Water Distribution System Modelling

1930's	1960's	1970's	Mid 1980's	Late 1980's	1990's	2000's	Future
Hardy Cross Network Flow Analysis	Computer Analysis of Networks	Widely available models for mainframes andminis	PC based models & Steady-state water quality models	Dynamic water quality models	Integrated modeling –mapping –database-GIS – SACDA & Containment kinetics & Optimization		Multi-platform models,Critical Analysis Management, Integration of TransparentGIS, Detailed water quality modeling

(Source: www.waterworld.com)

The rapid increase in population exerts a tremendous burden on the infrastructure of the cities. The government has planned to upgrade many cities in India to ensure the continuous supply through various schemes to improve and set infrastructural development for the coming future called Atal Mission of Renewal of Urban Towns (AMRUT) to convert 100 cities to urban cities (www.waterworld.com). Water is one of the fundamental necessities for sustaining life on the earth for all living entities. A human being needs water for many of his activities in day-to-day life. During the earlier era, most of the population on the earth used to settle beside nearby water bodies such as lakes, rivers. The water is supplied through closed conduits (pipe networks) to all the areas. The water distribution system consists of several components such as intake, pumping, transmission, treatment, storage, distribution network. A tremendous amount of money is spent every year worldwide to provide or upgrade drinking water facilities; nearly 80-85 % of the cost of a water supply project is used in the distribution system. Unheard-of increases in population, urbanization, agricultural expansion, and industrialization lead to higher human activities.

A developed urban city should have continuous water supply arrangement and should be capable of making adjustments in daily operations to ensure continuous water supply to improve service delivery; despite immense efforts, a large portion of the urban population, especially the poor, still lack poTable water supply (Chary, 2005). Even though a target of 140 liters per capita per day (LPCD) has been set, the average per capita water supply ranges between 57 to 160 LPCD. In the slum area, the average supply attains a lower value of a mere 27 LPCD, which is a matter of concern as it involves serious health risks. One of the important causes of poor service in many Indian cities is the intermittent water supply. Presently, no medium or large-size Indian water utility supplies 24x7 water to their customers. (indiawaterportal.org).

Global warming concerns and the worldwide increase in vulnerability events as floods and droughts have raised the issue of preserving and distributing water properly from accumulated areas to the demanding area within various restricting urbanization policies. The study was carried out in Navi Mumbai, Maharashtra, India, on a development site of rehabilitation. The technique of designing a sustainable Water Distribution Network (WDN) more precisely with the application Geographic Information System (GIS) and through well-mannered simulation is the aim of this study. The principle objective is by an alternative approach of designing an optimized WDN where flow in-network is through gravity, neglecting the use of pumps wherever possible, which supported in investigating the various hydraulic parameters like velocity, flow, head in the network. This study discussed an alternative approach in designing a sustainable WDN considering GIS. DEMs through satellite data, the elevations directly impacted the various hydraulic parameters such as velocity, the flow which played a vital role in gravity flow as well as the pipe diameter and its length was designed considering the street maps, which along with its roughness coefficient helped in optimizing the network to a reasonable extent. Investigating the various hydraulic parameters with the application of GIS in identifying the proper locations for placing the water tank and other nodes in WDN with an optimistic approach is the specialty of this research.

The fundamental objective of urban water service is to supply adequate and safe water efficiently and effectively at an affordable price. Intermittent water supply is defined as "a piped water supply system which delivers water to its users for less than 24 hours in a day." An intermittent system has many drawbacks. In non-supply hours, partial vacuum conditions are formed (Figure 1.1) inside pipelines due to water drawn from the pipe.

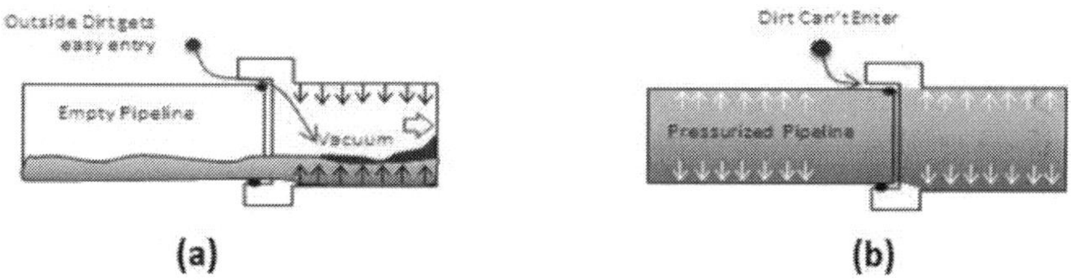

Figure 1.1: (a) Water is contaminated in an intermittent supply. (b) Dirt cannot enter water flows

In a continuous water supply, outside contaminants cannot enter in water due to internal pressure. This study would focus on the problems due to extreme rainfall, which accumulates and flows at street level. The European Flood Risk Directive [EU, 2007] promotes flood risk management plans with non-structural measures aiming at the resilience of urban infrastructures and preparedness of the social system. Drawbacks of the intermittent water supply system and the merits of the 24x7 system are shown in Table 1.1 and Figure 1.2.

Table 1.2: Drawbacks-Intermittent water supply system, merits of 24 x 7 system

Intermittent	24 x 7 Continuous
Pollution due to ingress of contaminants	Better health outcomes
Large doses of Chlorine	Fewer doses of Chlorine
Capacities underutilized	Tank capacities optimally utilized
Frequent daily operations - Valves suffer wear and tear	No daily operations-Steady pressures- Life of network increases
More human resources- Zoning	Less human resources
Larger sizes of pipes	Optimum sizes of pipes
Supply hours affect poor	Water is accessible to poor
Pay for pumping	Pumping costs are negligible
Meters go out of order	Meters work satisfactorily
Store and throw away tendency	Water reliance
Wastage of treated water	No such wastage
Less willingness to pay	Willingness to pay- even in slums
3 hours for flushing water	Time is spared for other rewarding activities.
No industrial growth	Attracts industries

(**Source:** www.dahasahasra.com)

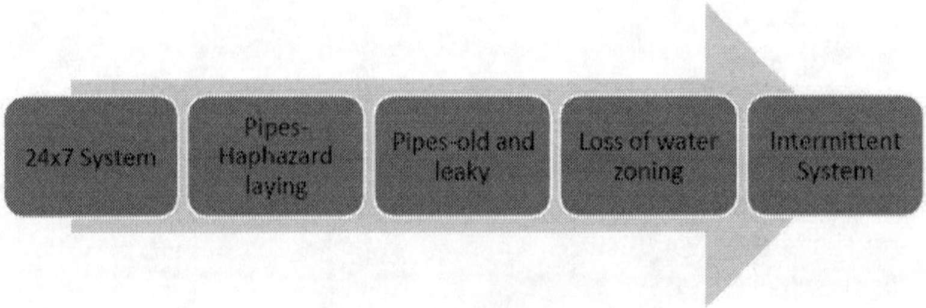

Figure 1.2: Undesired transformation of Continuous system to intermittent system

Within the field of water distribution networks (WDN) analysis, case study investigation could be a well-known instrument for testing or benchmarking advancement procedures, assessing recent created advances, demonstrating theories, testing modern computer programs, and examining instabilities in models. However, with a constrained number of cases ponders for assessments, examinations are case particular, and it is not easy to exchange the obtained results about other boundary conditions. In expansion, exceptionally contrasting case studies make it troublesome to distinguish driving factors. In any study, information accessibility for genuine case research is constrained due to the time and need for information collection. Besides, the information on WDN that is accessible is regularly fragmented or of inadequacy quality. Property of the hydraulic modeling tool is that it can be used to accurately simulate a network before it has been built or modified. While the simulation of the network, problems can be removed. Geem Z. W. et al. (2011).

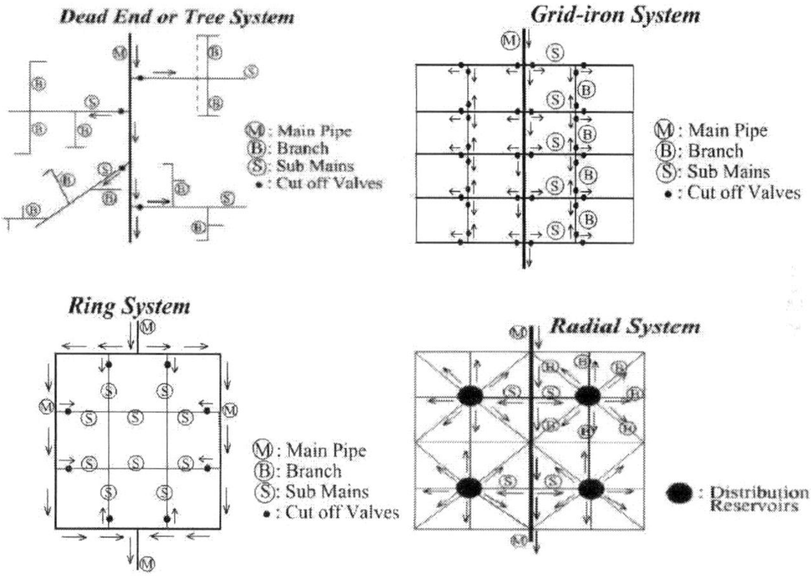

Figure 1.3: Dead end system, grid iron system, ring system, and radial system (**Source:** https://www.wateronline.com)

Terrorist attacks have inbuilt fear and restricted access to sensitive WDN data (Brumbelow et al., 2007). Be that as it may, such information is vital to investigate (Di Pierro et al., 2009), water quality investigations or appraisals (Rodriguez and Serodes, 1999; May et al., 2008; Tamminen et al., 2008; Wu et al., 2009a), execution examination (Ramos et al., 2009) and demand estimating had to utilize one or at most several WDS case considers as application examples. Work carried out by two loop frameworks has been discussed in the literature. First, it conveys water from the source(s), i.e., reservoir, water tank. Second, to the consumer(s), i.e., destination through the most excellent conceivable course would optimize the WDN, and thus its survival comprises of three components, i.e., the source of water, its treatment unit, and the distribution network. Water source shows rivers, reservoirs, wells, tanks. The treatment unit keeps up the quality as per guidelines before supply it towards its destination, and finally, the distribution network needs to be laid to meet its infrastructural and demand requirement. An appropriate well-outlined WDN framework comprises pipes, junctions, pumps, valves, tanks. It would be economical when it can supply the specified demand of water to the different commercial, household, and industrial stakeholders. Depending on the necessity of water utilization in different divisions, the slightest pressure should be directed as per the administering standards for WDN frameworks. The various layout of water distribution networks such as dead-end systems, grid iron systems, ring systems, radial systems is shown in Figure 1.3.

1.2 REMOTE SENSING

The science of collecting information of all the features or objects over the surface of the earth without coming in physical contact with them or acquiring information of any phenomenon instead of collecting data on-site is termed Remote Sensing This science, when connects with technology, is used in various applications across various field such as surveying, geography. In the past few decades, various disciplines in Earth science are using remote sensing in ecology, hydrology, oceanography, glaciology, and commercial, economic, urban planning, and military applications, to list a few. Aircraft or satellite-based sensor technology are used in Remote sensing to understand and classify the various features on the surface of the earth, oceans, atmosphere, using propagated electromagnetic radiations in terms of signals. The capability of satellite sensors to capture reflectance from land cover features in various spectral bands of the electromagnetic spectrum provides an exciting opportunity to study the behavior of various land cover features. The land cover features exhibit distinct characteristics in the different spectral ranges. These characteristics are reflected in the spectral response recorded by the sensors operating in the particular spectral band. Apart from the distinct spectral band, the satellite data is available in various periods as well. This temporal satellite data allows the study of change dynamics in land cover features.

1.2.1 Importance of Remote Sensing

'Remote sensing is the science of making inferences about objects from measurements, made at a distance, without coming in contact with the objects under study.' —*George Joseph*

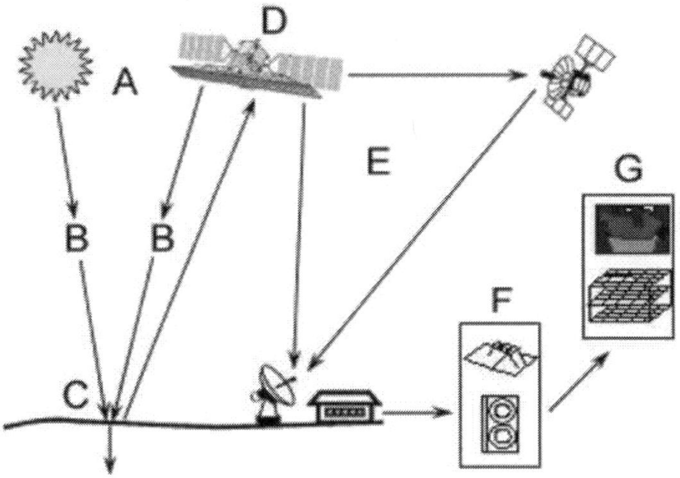

Figure 1.4: Components of Remote Sensing
(**Source:** https://theconstructor.org/others/remote-sensing-applications/398/)

As shown in Figure 1.4, A indicates the sun (A), which is the energy source of electromagnetic radiation (EMR), which passes through the atmosphere(B) and interacts with its obstructions such as haze, clouds, ., and hits the target on the ground at (C). Some EMR is absorbed, and others are emitted again through B, which later the sensor (D) on the satellite captures and records. This reflectance which is recorded is transmitted back, passing again through (E) to the ground station (F), where the data is processed in which errors are removed and converted to raster images. Later, the engineers and scientists interpret and do the necessary analysis for various application Flies, as mentioned above.

1.2.2 Electromagnetic Radiation (EMR)

In remote sensing, almost most of the energy uses electromagnetic energy, which is usually used in our daily lives. This energy through wavelength move ahead when it travels and is measured in frequency which denotes its speed. Electromagnetic energy can be explained on a broader scale known as an electromagnetic spectrum, as shown in Figure 1.5, and it varies from radio waves to gamma rays, i.e., higher to lower frequency, respectively.

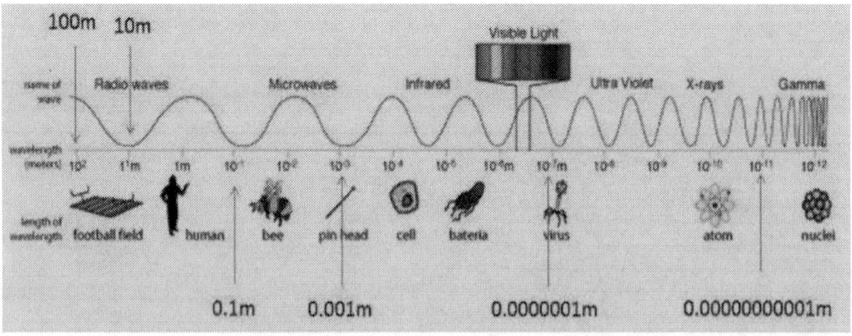

Figure 1.5: Electromagnetic Spectrum
(**Source:** https://i0.wp.com/berthabella.com)

1.2.3 Spectral Signatures

Spectral signatures are the characteristic curves associated with different surfaces (including types of vegetation) that show the extent to which the EMR of a particular wavelength is reflected. Thus, whereas the albedo (the proportion of the incident light or radiation that is reflected by a surface) tells you the proportion of total radiation reflected, the spectral signature tells us which wavelengths are either absorbed (a trough in the spectral signature) or reflected (a peak in the spectral signature), as shown in Figure 1.6.

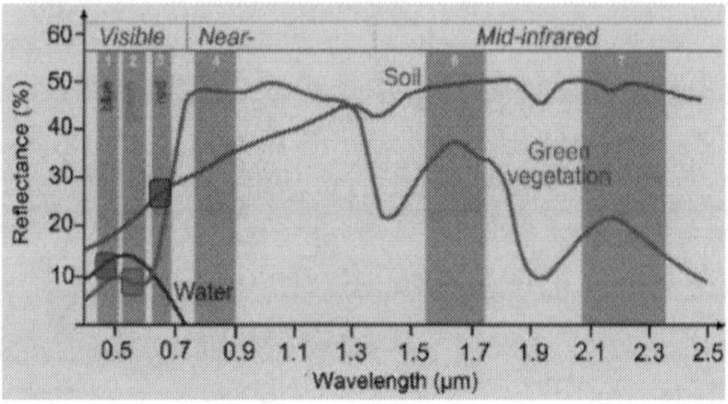

Figure 1.6: Spectral Signature

1.3 CONCEPTS OF GEOGRAPHIC INFORMATION SYSTEM (GIS)

GIS is more used as a tool in various applications, as stated above in today's advanced world where users have various interactive queries such as analyzing spatial information, editing (to add or to improve) attributes in terms of data in maps. Hence GIS is applied across various services that depend upon location intelligence applications wherein efficient and effective analysis and visualization. The Geographic Information System (GIS) is taken to visualize the sources and feeders and conceptualize the entire distribution network. K. Vairavamoorthy et al. (2007) presented the development of a new GIS-based tool that predicts the risks associated with contaminated water entering water distribution systems from surrounding foul water bodies such as sewers, drains, and ditches. Baiyi Jiang1 et al. (2012) considered Dynamic pipe network hydraulic modeling is the realization of the municipal water supply network to optimize an effective means of scheduling and science operations. A detailed analysis of the operation using Water GEMS provides practical examples to establish a water supply network model based on Supervisory Control and Data Acquisition (SCADA) data. A Saminu, Abubakar (2013) worked on GIS in spatial planning support tools has an essential advantage through changing the valuation standards to visually illustrate and depict where the implications of different spatial decisions and alternatives are convenient. I Soo Lee et al. (2014) used a geographic information system (GIS) to predict the spatial distribution of water demand according to building units by applying the basic unit of water use by purpose. Based on the results, the buildings were then grouped into blocks to produce a methodology for controlling small districts using a microscopic approach.

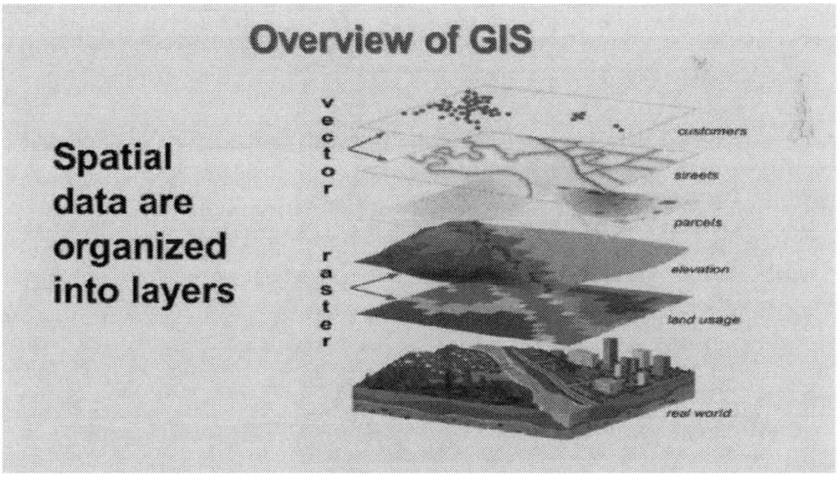

Figure 1.7: GIS Mapping

A Geographic Information System (GIS) can be defined as a system for entering, storing, manipulating, analyzing, and displaying geographic or spatial data, as shown in Figure 1.7. These data are represented by points, lines, and polygons and their associated attributes (i.e., characteristics of the features which the points, lines, and polygons represent).

1.3.1 Georeferencing

Georeferencing is the process of assigning spatial coordinates to data that is spatial in nature but has no explicit geographic coordinate system. (Refer to Figure 1.8)

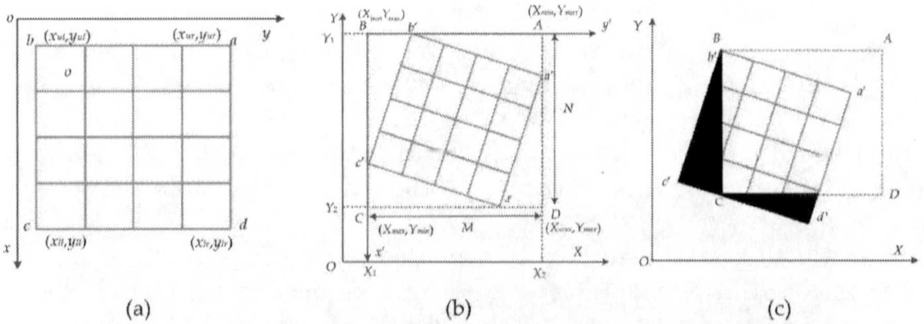

Figure 1.8: Geo Referencing
(**Source:** https://www.mdpi.com)

1.3.2 Importance of Digital Elevation Model (DEM)

Observing the earth's surface from space through satellite images and understanding the terrain profile helps in various applications such as constructing infrastructural projects such as dams, tunnels, irrigation channels. In the past few decades, many researchers have studied environmental aspects such as ecosystem, global warming, floods, disaster management, and water distribution networks and based on satellite images and given near to accurate results compared to in-situ survey analysis. Very few organizations worldwide provide free good resolution satellite datasets for research activities. Due to this, the research fraternity, especially students, even though well equipped with hardware and software, and technology, are handicapped to achieve the goals with limited resources. The study area lies in the Raigad district near Navi Mumbai, India, where the terrain varies. To overcome this aspect, the Indian Space Research Organisation (ISRO) has opened the Bhuvan portal for freely downloading satellite images. Hence, this research aims to create a False Color Composite (FCC) by image stacking process in Quantum GIS (QGIS), open-source software. These satellite images are Resoursesat-I (LISS-III) and Cartosat-I (DEM), freely downloaded from the Bhuvan portal. In this study, the FCC raster file is superimposed over the Cartosat-I satellite image to simulate the Digital Elevation Model (DEM) of the study area in 3D. For the Image stacking process, EPSG:3857-WGS 84/ Pseudo Mercator coordinate reference system is assigned in the methodology while developing a standard FCC. The results developed through image stacking produce output in TIFF format, and with the application of the QGIS2threejs plugin, the DEM 3D model of the study is simulated, giving an in-depth view of the terrain conditions of the study area. The results indicate a well-defined ground profile with latitude, longitude, and altitude of every point of the study area; hence would help in laying water distribution networks according to gravity the ground slopes

1.4 SUSTAINABLE WATER INFRASTRUCTURE

1.4.1 Introduction to Water Transmission Network (WTN)

Water Transmission Networks (WTNs) serve many purposes and provide water for human consumption. Piped water is used for washing, sanitation, irrigation, and firefighting. Networks are designed to meet peak demands; in parts of the network, this creates low-flow conditions that can contribute to microbial and chemical water quality deterioration. The purpose of a system of pipes is to supply water at adequate pressure and flow. However, the pressure is lost by the action of friction at the pipe wall. The pressure loss is also dependent on the water demand, pipe length, gradient, and diameter. Several established empirical equations describe the pressure-flow relationship (Webber, 1971), and these have been incorporated into network modelling software packages to facilitate their solution and use. However, there is still not a convenient evaluation for the reliability of water distribution systems. Traditionally, a water distribution network design is based on the proposed street plan and the topography. Using commercial software, the modeler simulates flows and pressures in the network and flows in and out to/from the tank for essential loadings.

1.4.2 Introduction to Water Distribution Network (WDN)

Water Distribution Network (WDN) comprises three primary components; water source, treatment, and distribution network. Water distribution systems serve the community and help power the economy by delivering water from a source(s) to its consumers. Water sources can be reservoirs, rivers, and groundwater wells. Water treatment facilities disinfect the water to drinking water quality standards before delivering it to its consumers. The distribution network is responsible for delivering water from the source or treatment facilities to its consumers at serviceable pressures and mainly consists of pipes, pumps, junctions (nodes), valves, fittings, and storage tanks. WDS is required to supply water to domestic, commercial, and industrial entities above or at a threshold pressure with consumer demands that vary throughout the day, week, season, and year. The minimum pressure observed at junctions throughout the system varies depending on the water-consuming sector and regulations governing the distribution system.

This study summarizes the work carried out by various researchers on the analysis and optimization of water distribution networks & their effectiveness in the water distribution industry. The study's objective is to determine various parameters in the water distribution network of the study area, such as discharge, velocity, pressure, heat loss. Controls pipe design, i.e., diameter and merits, different pipe materials would be assigned, such as galvanized iron, cast iron steel, concrete. Furthermore, its adherence to the design parameters would be compared. For a more realistic approach, the water distribution network is simulated for a single period analysis, i.e. (24 hours run with a demand pattern). Keeping the junctions and pipes as per city map and assigning roughness coefficient for different pipe materials used various iterations is done and based on the result obtained the most efficient and effective pipe material can be finalized to make the network more sustainable.

Ingeduld P. et al. (2006) studied the intermittent water supply scheme with two case studies of Shillong, India, and Dhaka, Bangladesh using the EPANET tool. The alternate emptying and refilling of water pipelines make it challenging to apply standard EPANET based hydraulic models because of low pressure and empty pipes. There are various layouts of water distribution networks such as the Dead End System, Grid Iron System, Ring System, and Radial System. EPANET is a computer program that performs an extended period simulation of hydraulic and water quality behaviour within pressurized pipe networks. EPANET tracks the flow of water in each pipe, the pressure at each node, the height of water in each tank, and the concentration of a chemical species throughout the network during a simulation period comprised of multiple time steps. In addition to chemical species, water quality and source tracing can also be simulated. EPANET is designed to be a research tool for improving our understanding of the movement and fate of drinking water constituents within distribution systems. It can be used for many different kinds of applications in distribution systems analysis.

1.4.3 Introduction to Water Quality

The integrity of well-managed distribution systems is one of the most significant barriers that protect drinking water from contamination. However, the management of distribution systems often receives too little attention. Distribution systems can incorrectly be viewed as passive systems, with the only requirement being to transport drinking water from the outlets of treatment plants to consumers. There is extensive evidence that inadequate drinking-water distribution systems have led to illness outbreaks in developed and developing countries. The causes of these outbreaks and the range of chemical and microbial hazards involved are diverse. The shortage of water and the deficiency of safe drinking water is one of the most challenging issues of this century. It consists of a complex network of pipelines, pumps, tanks, water reclamation facilities, and treatment stations. Thus sources of water and treatments have a significant effect on the quality of the distributed water. Contamination of water occurs by many mechanisms & pathways (Islam et al. 2015).

The most predominantly used water treatment disinfectant is chlorine. Chlorine is usually injected after all treatments at a water treatment plant to disinfect poTable water and maintain a residual within a water distribution system (WDS), preventing the regrowth of pathogenic bacteria. Moreover, chlorine is a chemical decay over time in the distribution system, and thus the amount of chlorine might vary from one point to the other. Hence a higher dose of chlorine is added to the treated water before it enters the distribution system. However, this high dose of chlorine generally results in taste and odor complaints from consumers residing in localities closer to the source. Furthermore, it might react with natural organic matter to form by-products, some of which may be carcinogens. Maintaining an optimum level of residual chlorine and reducing the detrimental effects caused by the additional chlorine dosage is a challenging feat for the concerned authorities. A potential solution to mitigate these problems is booster chlorination, in which required doses of chlorine are added at specific locations at which the amount of residual chlorine reduces below the minimum specified amount (0.2 mg/l). This method helps the authorities in maintaining the required maximum and minimum chlorine concentration. Chlorination dosages are typically employed in the waterworks against certain parasites. They can be excluded only by effective filtration or by higher chlorination doses that can be tolerated without subsequent de-chlorination. In practice, more significant than 0.5mg of free chlorine for 1 hr is sufficient to inactivate the virus. Even in water that was initially being polluted, the water is free from particulates and organic matter.

1.4.4 Mobile Application

Nowadays, data collection has been considered the most expensive and time-consuming in a GIS project. It can cost up to 85 % of the total cost of a project. Probably due to the high cost of the equipment needed for data capturing and the labor cost. For example, collecting the coordinate of a point using the GPS surveying method would take at least fifteen minutes to be completed. What if there are a hundred points to be captured? It might take days with the conventional survey methods due to the time wasted setting up the equipment and transferring it to other points. The same procedure can be repeated using ODK (Open Data Kit), a free, open-source suite of tools that allows data collection using Android mobile devices and data submission to an online server, even without an Internet connection or mobile carrier service the time of data collection. Collect is an open-source application that is implemented on mobile smartphones. It is part of a suite of tools developed by a group at the University of Washington called Change, which explores technology in improving lives in developing countries.

The cloud server has been set up in the Kobo toolbox created manually. The required form is built into this server and transfers the uploaded form to the ODK application using this server URL, username, and password. The data is collected from this application and co-ordinates of different location is collected with its picture. This application also supports incorporating GPS points, photos, videos, bar codes, and sound bites.

The water transmission line has been located from this survey and alignment of the existing pipeline. Mobile data collection or mobile surveys is an increasingly popular method of data collection. Over 50 % of surveys today are opened on mobile devices. The survey, form, app, or collection tool is on a mobile device such as a smartphone or a Tablet. These devices offer innovative ways to gather data and eliminate the laborious "data entry" (of paper form data into a computer), delaying data analysis and understanding. By eliminating paper, mobile data collection can also dramatically reduce costs: one World Bank study in Guatemala found a 71% decrease in cost while using mobile data collection. A more no. of functions is available in this application, and it is easy to use for all people. The reduced level was also collected in this application photos, videos, and text was also collected. The data can be easily stored in the ODK server and exported in many formats. The data can be used for the QGIS application and Google Earth.

1.5 SCOPE, GAPS, NECESSITY AND METHODS

The various research study, technical reviews, journal, and conference proceedings papers referred to in this literature study have shown a direction for this research. These directions were classified in terms of the below-mentioned parameters such as scopes, gaps, necessity and methods. Based on which this research study's objective is formed, these parameters helped give this research a backbone by which this study can see a standstill. Several papers have appeared on the subject of analysis and optimal water supply networks consisting of several sources, pipes, valves, pumps, and reservoirs in the recent past. These analyses usually need for an efficient technique to compute the pressure and flows in a defined network at defined levels of computational accuracy, given the characteristics of the network such as the pipe lengths, diameters, and friction factors as well as the hydraulic characteristics of all ancillary plant such as pumps and valves. Most network simulation models used in current engineering practice are based on the conventional Demand Driven Simulation Method (DDSM). They assume that nodal outflows are fixed and are satisfied regardless of network pressures. The assumption simplifies the mathematical solution of the problem but is not always appropriate because it is clear that the amount of outflow at nodal outlets depends on network pressures. The flow would be significantly reduced if the pressure falls below a minimum required level (due to some critical events such as mechanical and hydraulic failures or excess demand). Although some nodes may satisfy their demands, others may meet the demand partially, while the rest may fail and may not provide any water at all. Therefore, the assumption of fixed nodal consumptions is valid only under normal conditions when the pressures can be expected to be adequate to satisfy the stipulated demands.

1.5.1 Sustainable Water Distribution Network

This session would discuss in detail the various parameters which are being investigated in this research.

Sub Title	Parameters	Work Area
Sustainable water distribution network	Velocity, pressure along with flow and head loss	Water Distribution Network (WDN)
	Pipe diameter & material considering pipe roughness co-eff	Water Transmission Network (WTN)

The various scope in this field are that Planners and engineers attempting to improve the resilience of water distribution systems. Water distribution systems may be represented as network efficiency and robustness to failures. Three-dimensional (3D) visualization (here my study would use shapes files using GIS).

Digital Elevation Model (DEM) in laying the pipes By observing the gaps in the Literature survey, studies can be done in increasing water use efficiency, productivity and to quantify the economic, social aspects to make WDN and WTN more sustainable Drawbacks of the intermittent water supply system and the inability to shift to continuous supply mode are the main challenges in developing countries, especially India. In addition, the suitability of the infrastructure laid over the past two to three decades to meet the 24/7 demand of today's population is an issue for many water managers. The present study addresses this issue using EPANET *software for a pilot study and QGIS* to generate hydraulic simulation and shape files.

1.5.2 Water quality

This session would discuss in detail the various parameters which are being investigated in this research.

Sub Title	Parameters	Work Area
Quality	Chlorine decay	Providing Booster stations

In recent years, contamination of water in water distribution systems (WDS) has been recognized as a significant cause of waterborne diseases. Owing to old and deteriorated pipes and the presence of pollution sources in contact with the supply line, water boards often decide to replace the network in the interests of public health. GIS maps can be prepared for the water supply network . Studies in-depth can be done: (gaps observed in Literature survey)

Chlorine decay in a water distribution system can occur because of reactions within the bulk fluid (bulk decay) and from reactions with materials associated with the pipe wall (wall decay) (Vasconcelos, 1997). Bulk decay coefficients for chlorine depend on the nature of the source water and the treatment it has received. Wall decay coefficients depend on the pipe material and its condition (e.g., biofilm growth or corrosion rate).

There is currently no established method for directly determining the kinetics of chlorine decay attribuTable to pipe wall reactions, and it is recommended to calibrate the model using the field data for determining the appropriate wall decay. The chlorine calibration process can be done in two main phases. The first phase would involve providing chlorine boosters at the source The second phase would involve providing chlorine boosters at a source + "n" nodes (here, various trial and error methods need to research, and finally, the best output can be generated by plotting graphs for residual chlorine.

1.5.3 Security Management

This session would discuss in detail the various parameters which are being investigated in this research.

Sub Title	Parameters	Work Area
Security Management	Leakage, maintenance, Non-Revenue Water (NRW)	Detecting techniques, Mobile apps

Due to high pressure and velocity in WTN and WDN, there is always a high risk of a damage- causing scenario in various values, nodes. Throughout the water supply network, which remains unchecked many times is compounded by the age of the infrastructure

A touch screen technique to find out the loopholes anywhere in the water supply network Water leakage, maintenance, and NRW in WTN/ WDN are severe problems for many cities and a massive challenge for water utilities. An integrated system for the detection, early warning, and control of these gaps has to be developed and successfully used to manage the pipeline networks. In this approach, a mobile application through Open Data kit ODK is designed wherein issues such as leakage, maintenances, and NRW from reported and un-reported bursts and background leakage can be known soon with accurate or estimated indices of leakage performance are evaluated.

1.5.4 Application of Remote Sensing and GIS

This session would discuss in detail the various parameters which are being investigated in this research.

Sub Title	Parameters	Work Area
Application of remote sensing GIS	Satellite image and LULC maps, False Color Composite	Image Stacking, Shape Files, Geo refereeing DEM

A bibliometric analysis of published geographical information system (GIS) research was performed to evaluate current research trends, quantitatively and qualitatively. The results showed that the growth of scientific outputs has exploded since 1991, with an increasing collaboration index, references, and citations. Environmental sciences, multidisciplinary geosciences, ecology, physical geography, water resources, geography, and remote sensing To organize a wide range of data to be used in decision-making models, including the plans aimed for pipe replacement. Digital elevation models and shapes files are not prepared for my study area, which can be used in the study of WDN and WTN.

Planning, management, maintenance, and emergency-response of pipeline network require a series of professional analysis tools.

From this perspective, GIS provides necessary technical support for the pipeline network analysis. This platform can provide a scientific basis for network planning and reconstruction, improve the efficiency of emergency responses, and advance the level of daily operations such as inspection and maintenance Using the data provided from an entire network, the advantages of the proposed methodology are evaluated by using various satellite images via the Bhuvan portal. Satellite images can be downloaded as per required spatial, spectral, and temporal resolutions and can be stacked, and digital image processing can be done in QGIS and other software.

1.6 OBJECTIVES

The objective is to design safe water networks using the hydraulic modeling tool and GIS. The study of existing water distribution networks and O&M systems can help identify the loopholes in the system and incorporate the various parameters in the optimization model for the water supply network design. Based on the various literature review and gaps observed in them as mentioned in the earlier chapter following objectives for the present studies have been arrived: There are four primary objectives of this research

1. To prepare Land Use Land Cover (LULC) maps of the study area by Remote Sensing Techniques.
    - To Geo-reference the study area with toposheet in QGIS
    - To generate 3D simulation Digital Elevation Model (DEM) in QGIS using CartoSat and ResourseSat Satellite Images.
    - To analysis Normalized Difference Vegetation Index (NDVI) and Enhanced Vegetative Index (EVI) Dynamics using LandSat Temporal Satellite data of 1998, 2009, and 2020 with 11 years intervals.
2. Carry out the various parametric Investigations by hydraulic approach.
    - To find the best possible Water Transmission Network (WTN) route by comparing eight possible causes.
    - To design the gravity-driven Water Distribution Network (WDN).
    - To provide chlorine booster station in WDN for Water Quality Analysis
3. Justification Techniques using Geographic Information System (GIS).

- By Satellite Navigation (Car Antenna)
- Mobile Application (ODK) for Security Management and data collection

4. USP: The uniqueness of this research is that the whole work is done by
    - Free Open-source GIS software (QGIS)
    - Free Open-source Hydraulic software (EPANET)
    - Free Satellite Data (ISRO's Bhuvan Portal)
    - Free Mobile Application platform (Open Data Kit-ODK)

1.7 PROBLEM STATEMENT AND STUDY AREA

As this research accentuation, the application of Remote sensing and GIS and its impedances with WTN and WDN, the region of interest is chosen considering the geographic and hydraulic perspectives. The study area is near Navi Mumbai as Figure 2, which lies in between 18°.58'.44.61" to 19°.0'.57.16" latitude and 73°.02'.54" to 73°.05'.39.61" longitude (ref Figure 1.9) in Maharashtra, India on a development site of rehabilitation.

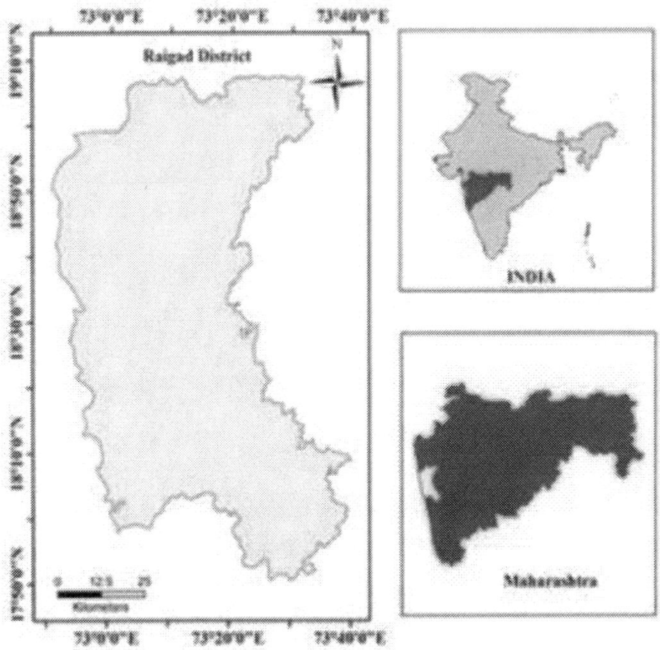

Figure 1.9: Study Area – Raigad, Maharashtra (India)

The WTN is laid between the source (Hetwane Dam) and the destination (Kunde Vahal) around 18.8 Km. The WDN is near a hillock near Kunde Vahal, the upcoming Navi Mumbai Airport. This study involves locating the most appropriate water tank position over the hilltop with DEM application.

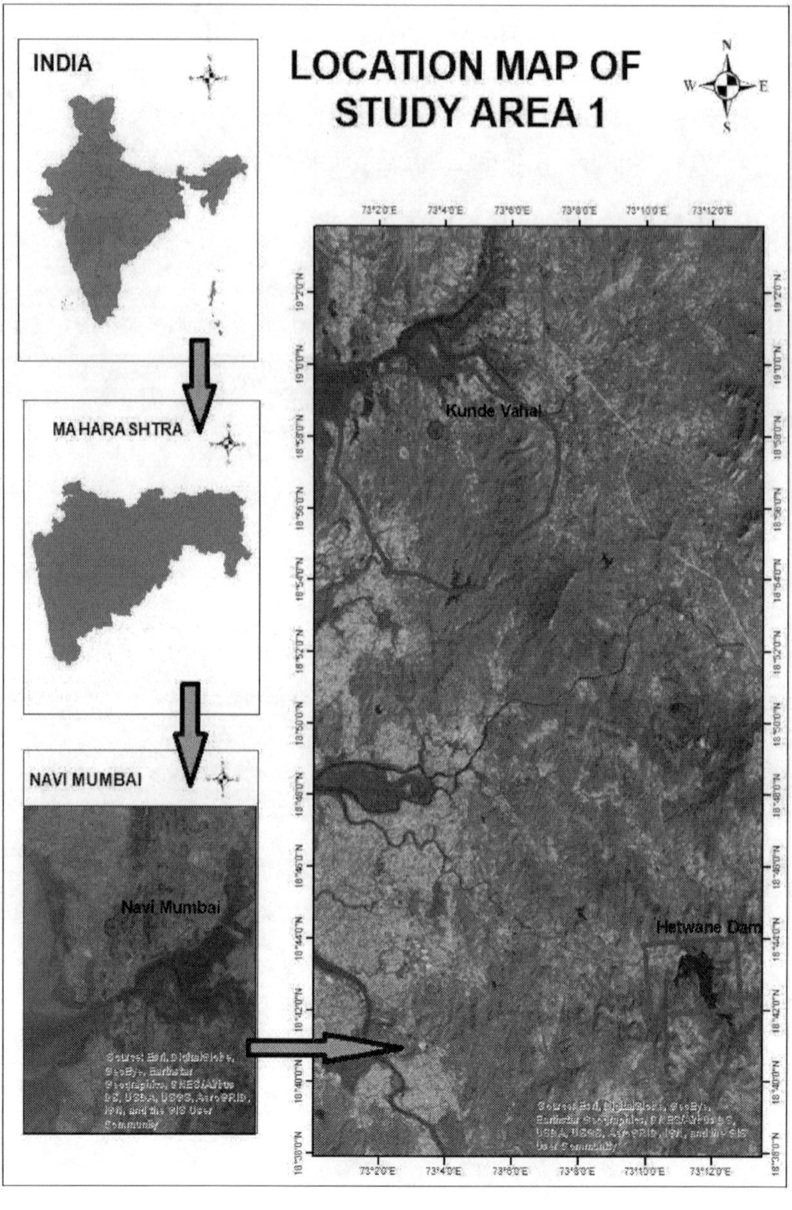

Figure 1.10: Location Map of the Study Area

1.8 STUDY AREA

This research explores the importance of Land use and land cover in urban areas and their boundaries. This study has considered the Raigad region very close to Mumbai and Navi Mumbai in Maharashtra, which is upcoming connectivity to many cities across India.The study region lies in between latitude (17°.50'N - 19°.10N') latitude and longitude (73°.00E - 73°.40E') as shown in Figure 1.10

The existing WTN is a PSC pipeline of a total 18.80 km length, which is having 1500mm diameter pipe for the first 11.90km and 1200 mm diameter for the remaining 6.90 km length in which the water is transmitted from the Hetwane dam to the sump at WTP, Jite as shown in Figure 1.11 and along with this 20 MLD of water is to 35 en route village nodes along with the existing 150 MLD which combines as 170 MLD.

Figure 1.11: Study Area showing Hetwane Dam in plan

An intake structure, allied works, and CIDCO tunnel are constructed to draw water from the Hetwane dam. The Reduced Levels (RLs) of outlets of the Hetawane dam are 50.50 m and 53.50 m with a residual head of 4 m. The elevation of WTP and raw water sump is at 36 m and 14 m, respectively. Figure 1.12 shows the WTN route from the source to the destination.

Figure 1.12: Water Transmission route from the source till destination (**Source:** Google Earth)

Water transmission systems may be represented as networks of multiple nodes, e.g., reservoirs, water treatment plants, storage tanks, and hydraulic junctions interconnected by physical links, e.g., pipes. We consider all the factors and design criteria such as pressure, velocity, discharge, friction factor, frictional losses, losses at bends, pipe materials, costs, junctions, different valves. As discussed earlier in the abstract, we are giving eight alternative alignments (Figure 1.13) and comparing them to give the most feasible transmission line, which would potentially meet the public water demands. We are also eliminating the surveys required for these kinds of projects with the help of modern software and techniques 8 Cases were analyzed in this research to select the best possible route of WTN.

This study involves locating the most excellent location for placing the water tank over the hilltop with DEM application. Figure 1.14 shows L-Section showing the Ground profile from the source to destination and laying the WDN down the hill slope, taking advantage of gravity without utilizing pumps corresponded with the study's objective in conjunction with optimization. The fast-emerging spatial information technology, remote sensing, GIS, and GPS are practical tools to overcome most

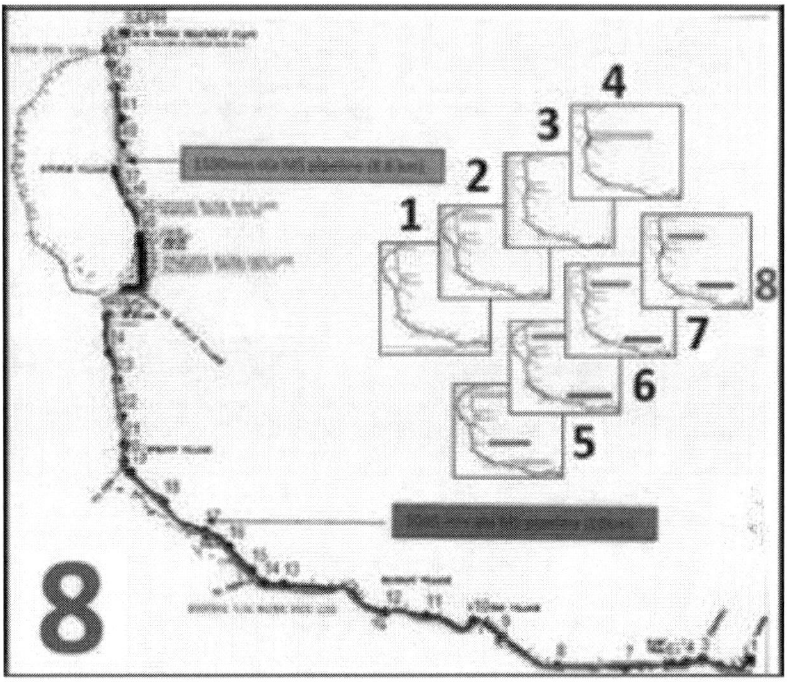

Figure 1.13: 8 Possible routes for Water Transmission Network route

land and water resources planning problems. Management instead of conventional strategies of data process Utilizing the LPS strategy for sensor geometry is modeling the extraction of the corresponding DEM created great results that are reasonable for the operational utilization in arranging and improving characteristic watersheds. The DEM accuracy analyzed both at the point mode, and the surface mode delivered great comes. Cartosat-1 stereo data can be considered as high accuracy.

1.9 LAYOUT OF THE THESIS

The thesis is divided into five major chapters, as given below.

Chapter 1 provides basic information about the introduction to Water Transmission Network (WTN) Water Distribution Network (WDN) with their water quality requirements. This chapter also emphasizes the importance of the mobile application technique to collect data. The various scopes, gaps, necessary and methods that were Figured out during the literature survey, and the basic hydraulic investigating parameters required to carry out the reliability analysis are detailed. The objective, problem statement, and study area details are explained

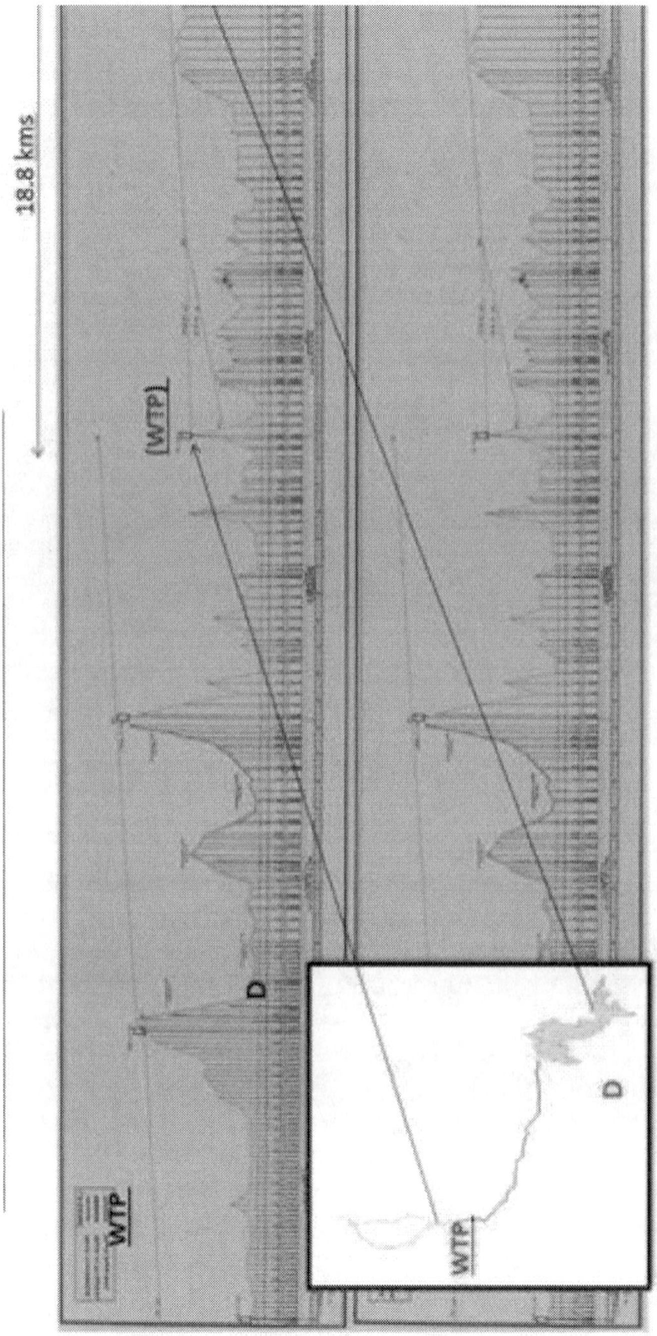

Figure 1.14: L-section showing the Ground profile from the source to destination (**Source:** CIDCO)

for which this research is considered. The objective of this research, based on the gaps the research interest which narrowed down the literature gaps to research objectives are stated down. The first is primarily to prepare the Land Use and Land Cover Maps (LULC) of the study area, whereas the hydraulic approach is explained in the second objective, and the third Spatial analysis through remote sensing and GIS is discussed in detail. Finally, the problem statement of this research, wherein the location map of the study area and its geographical parameters are shown to understand the need for the above-said objective for this research.

In **Chapter 2**, a critical literature review in the study area of water distribution and transmission network and water quality. The methods available for simulation of water networks such as EPANET, WATERGEMS. are also discussed in this chapter. Literature specifically from the past two decades was referred to more as this research deals with the latest technology dealing with remote sensing, GIS, and mobile applications. Many techniques, methods, simulations which the past researchers have carried out are studied carefully to design a unique template for this specific research.

In **Chapter 3**, the various methodology is explained in brief related to the whole research aspect in terms of LULC, water transmission and distribution network, water quality, and mobile applications in schematic and tabular format, which is easy to interpret. This chapter starts with data collection and system requirements that deal with downloading the satellite data and the software and hardware configuration. Then, a further step-by-step technique for analyzing various spatial and non-spatial calculations is explained in brief. Justification techniques are also included toward the end of this chapter.

Chapter 4 delivers the results and discussion obtained for the given objectives as stated in chapter 1. For WTN, WDN, and water quality, the analysis is carried in open source software such as Planet and QGIS, and the output is shown in a tabular and graphical format to provide a sound representation. The simulation model is cross-verified by two justification techniques, one by generating a mobile application and the other by Global Positioning System GPS. Land use and land cover maps are also generated wherein the various vegetative indices such as NDVI, EVI are calculated using Landsat images from the past two decades and good use of remote sensing and GIS outs.

Chapter 5 finally deals with the conclusion and future scope, wherein the results of every objective of this research are explained in detail and validated as per the said requirements. During these four years of research, the researcher talks about the future scope, which has been seen as a wide window to perform more research in hydraulics, remote sensing, and GIS.

Chapter - 02

Literature Review

2.1 GENERAL

As this research emphasis on water transmission and distribution network simulation with an application of latest technological concepts using remote sensing, GIS and mobile application the literature review in this chapter is classified in four primary class as given below:

1. Land Use Land Cover (LULC)
    - Georeferencing
    - Remote Sensing and GIS
2. Sustainable water supply network
    - Water Transmission Network
    - Water Distribution Network
    - Water Quality
3. Mobile Application and other related technology

This review has not only eases this research to understand this topic from a bird's eye point of view but also has gone through various research papers in and around the same research areas in the past three decades to find out the various gaps, scopes, necessities required to be developed in this arena of sustainable development of water distributional infrastructural networks.

2.2 LAND USE LAND COVER (LULC)

2.2.1 General

In the recent past, many new exciting developments can be seen for observing changes in land surfaces by applying time-series data in remote sensing of analysis. Especially for more accepTable spatial resolutions, LandSat images are more in demand as it is available through opensource and provides a wide range for research in various aspects. A vast history of time series data sets is used in the past to understand the change detection, and a fantastic review of various methods are listed out in 1981 (Singh,1989) and good other research of change detection can be seen in (Jensen,2005) and (Coppin et al., 2010). A VHRR Very High-Resolution Radiometers were used earlier for time series data by various researchers such as (DeFeries and Hansen, 1995) (Viovy et al., 1992) (Loveland et al., 1995). In the recent past, MODIS imagery has been quite common in analyzing change detection (refer Boschetti et al., 2009; Verbesselt et al., 2010; Friedl et al., 2002). Landsat data set for time series lagged MODIS as MODIS data set is quite ahead in various parameters including more acceptable resolution sensors. A well-organized dataset can be downloaded for analysis in time series as it deletes the effect of cloud cover and acquires binned to standard grids. (Justice et al., 2002). Atmospheric correction and cloud and cloud shadow masking improvements can be observed (Wulder et al., 2012).

2.2.2 Georeferencing

Panaskar S et al. (2019) have used the Landsat series dataset of 1998-2008 to find land use and land cover of goffered sheets of western ghats using GIS software, whereas Vivek Kumar et al. (2019) have used georeffered data for waster budget assessment using GIS as SWAT tool for lower Godavari basin in Maharashtra. K Nagarajan et al. (2017,2018) presented their research on water distribution systems in the Raigad area using QGIS software as the primary tool to understand the ground slope for laying pipelines using the Digital Elevation Model. Priyanka Bhatkar et al. (2019) have worked on oil spills effects over the ecosystem of coastal areas in the Alibaug region in Raigad district in Maharashtra using data of pollution due to tarballs. Shobana Jadhav et al. (2019) have comprehensively used QGIS to find the best possible shorted route in the construction industry vehicular traffic transportation system in Mumbai with shapefiles, top sheet, and satellite data. Aditya S et al. (2019) have used Arc GIS to understand the LULC planning with an integrated approach and classified areas with vacant space to highly used congested areas for infrastructural development Akola Taluka Maharashtra. Sunilkumar Patel et al. (2019) used GIS to find peak runoff of stormwater in the H-East ward of Mumbai using georeffered toposheet and implemented them for his research to use the rooftop water to be utilized for necessary purposes using blue roof system. Chowdary Mohanlal et al. (2019) have applied a 4D GIS model in construction management in the Uran area, which is in the Raigad region using QGIS and has simulated its output for better visual representation. Reshma Kamble et al. (2019) have analyzed and designed a water distribution network using toposheet and georeffered Sawale Village in Raigad region in the GIS platform.

2.2.3 Remote Sensing and GIS

K. Vairavamoorthy et al. (2007) this study presents the development of a new GIS-based tool that predicts the risks associated with contaminated water entering water distribution systems from surrounding foul water bodies such as sewers, drains, and ditches. Baiyi Jiang et al. (2012). Dynamic pipe network hydraulic modeling is the realization of the municipal water supply network to optimize an effective means of scheduling and science operations. A detailed analysis of the operation using WaterGEMS provides practical examples to establish a water supply network model based on SCADA data. Ji Soo Lee et al. (2014) This study used a geographic information system (GIS) to predict the spatial distribution of water demand according to building units by applying the basic unit of water use by purpose. Based on the results, the buildings were then grouped into blocks to produce a methodology for controlling small districts using a microscopic approach A Saminu, Abubakar (2013). The application of GIS in spatial planning support tools has an essential advantage through changing the valuation standards to visually illustrate and depict where the implications of different spatial decisions and alternatives are convenient.

2.3 SUSTAINABLE WATER SUPPLY NETWORK

2.3.1 General

Jolly M et al. (2014) describes developing a database of several water distribution systems synthesized from a state-wide database of systems. The developed models include both small and medium networks, as well a range of system characteristics (i.e., system configuration grid, looped, and branched systems and system components number of tanks, pump Thomas Walski et al. (2015), This study provides an insight into the selection of the most suiTable configuration for closed distribution systems (i.e., systems with no storage capacity). The analysis is developed considering a wide range of operational scenarios in terms of flows and required heads. For each scenario, various design configurations are considered and compared.

C. Sou-Sen Leu et al. (2016) This study presents a novel model using expert structural expectation-maximization for predicting water leakage in WDS. Predicting water leakage is one of the most crucial tasks in water resource management; however, it is also one of the most challenging ones. A. Antonowicz1 et al. (2018) displayed a strategy of utilizing EPANET solver to back oversee water distribution system in Smart City by permitting to perform both single and cyclic simulations. Jiang B et al. (2012) expressed that dynamic pipe network hydraulic modeling is the realization of the municipal water supply network to optimize an effective means of scheduling and operations.

2.3.2 Literature Review of Water Transmission Network

H Ramesh et al. (2011) cited that a water supply system is a system of engineered hydrologic and hydraulic components that provide water supply. The major objective was to generate satellite-based thematic layers, town and ward boundary maps, and Geospatial Information System-based census data and estimate water demand, design of transmission lines, and main pipelines to meet future demand requirements. GIS has been used to integrate and estimate the quantity of earthwork to be excavated to cut and fill through the Digital Elevation Model (DEM). The pipe network system is simulated to understand its behavior for different inputs using EPANET 2.0. In the present study, both single period and extended period simulations were carried out for one ward's distribution network system. Simulation has been carried out for hydraulic parameters such as heat, pressure, and flow rate. Total Energy Line and Hydraulic Gradient Line were prepared for the simulated results.

Arunkumar M. et al. (2011) examined the water demand analysis of Public Water Supply in Municipalities using EPANET 2.0 software to provide effective planning, development, and operation of water supply and distribution networks, one of the essential components of urban infrastructure. Several factors ranging from population expansion to inadequate existing facilities are thought to be responsible for the frequent shortage in the water supply to the metropolis. The study delineated the areas within the municipality that are unserved or underserved by the Municipality. Using EPANET 2.0 water model software, the demand for the Underserved and Unserved area is calculated. A framework for taking management decisions such as extending the supply network and location of new facilities was given. David Ndegwa Kuria et al. (2012) In this work, a comprehensive solution for the Ganga Water Trust (GWT) has been developed to assist it in managing its water supply and distribution. The solution comprises two subcomponents: a mapping component and a billing component which are tightly coupled together. The proposed system uses sTable open-source products for the mapping component and the database. The GWT uses outdated maps and skies for the design and installation of a new water supply infrastructure. A billing system is in place which is used to manage client accounts, record meter readings, prepare bills, and record payments made. This presents a somewhat disjointed approach to the management of the water supply and its attendant infrastructure. The database that stores the account information is very different (softcopy) from storing the spatial information (hardcopy/paper-based). In the proposed solution, a single database is used, centralized, or distributed. The mapping component provides an interface through which the preliminary design of new and planned infrastructure can be done. After installation, these are reflected in the database, and the information becomes available as soon as it is stored. The billing component uses the same database to manage account information. Since the information is managed in one system, there is a streamlined and orderly flow from data collection to the final products from the system. The proposed solution leverages advancement in technology by providing two approaches - a desktop application for users within the Trust's intranet and a web mapping application for users utilizing the wider internet.

Michael Saadeh et al. (2013) The safety of drinking water supply is critical to the quality of life, people's health, and public welfare. However, the water main break is becoming more and more frequent and acute. These breaks are very costly to repair and cause many problems, including but not limited to flooding, blocking traffic, wasting water resources, and contaminating drinking water. Due to a limited budget, water utilities must use their resources wisely and proactively by prioritizing the pipe-fixing work. Therefore, it is imperative to replace the deteriorated old system pipelines with a new efficient and cost-effective design. The main objective of this project was to optimize the system and reduce the risk of water main breaks. The senior design project was implemented by re-designing a selected water-distributed network system starting from the Westside pump station on Illinois Road down to Washington Road in Western Fort Wayne, where more water main breaks have occurred in the last three years than the rest areas of Fort Wayne. The final design of this project consists of a redesigned approximately 7,300 ft. pipe network, a water tower, and galvanic corrosion protection. KY Pipe modeling was conducted to analyze and optimize the pressure and flow distributions in the pipe network. The new system would significantly save the cost of water main breaks and the energy consumption of the pumps. Jung, D., Kang et al.(2015) this study Develops a real-time near-optimal pump scheduling model for water transmission networks. The model aims to minimize energy cost utilizing optimal pump operation while meeting system demands and satisfying operational constraints. The real-time control is realized using Supervisory Control and Data Acquisition (SCADA) system, by which system monitoring data is collected, and the optimal control decisions are sent to the pumping stations for implementation. The model comprises three simulation modules, such as demand forecasting, hydraulic simulation, and optimization. Decision variables include pump on/off (binary code of 0/1) schedules for fixed-rate pumps for each time-step over a pre-defined operation time horizon. Various operational constraints are embedded in the model, reflecting real operational practices. The proposed optimization model is demonstrated using a simple, medium-sized water transmission system in Seoul, South Korea, and showed 19~27 % energy cost savings compared to a conventional operation scheme. Darshan Mehta et al. (2015) Pipe network analysis is the fluid flow through a hydraulics network containing several inter- connected branches whose aim is to determine the flow rates and pressure drops in the individual sections of the network. The classical approach for automatically solving the network problems is by using specialized software such as EPANET. This study aims to assess the performance of the Limbayat zone water supply scheme using hydraulic simulation software and address any improvements required to existing infrastructure and the mode of operation to improve the quantity and quality of water distributed to the consumers. The Limbayat zone is facing a water scarcity problem. The reason behind this scarcity is the pressure fluctuation. There is a large variation in the pressure head, and the pressure supplied is not sufficient to fulfill the requirement of water demand of the Limbayat zone. There may be leakages in the pipes, resulting in the pressure difference, resulting in water scarcity.

R. K. Rai, R. S. Tetu (June 2016) Describes how Applying mathematical models to obtain the optimal solution for water transmission systems has been of major concern during the last few decades. The optimization problem for the water transmission system is considered a non- linear problem during the search process to identify the decision variables (i.e., pipe diameter and cost of pipe and pump). In this study, the flow path algorithm has been adopted to obtain the optimal design of the basic unit of water transmission system involving pumping, i.e., three reservoir systems. The effect of pumping head on the optimal design of water transmission system has been considered in this work. B. A. Konnur, R. K. Rai (June 2016) Describes how The primary objectives of the National Water Policy of India 2002 are to improve the performance and cost-effectiveness of programs in the field of drinking water supply and to ensure the availability of an adequate quantity of drinking water of accepTable quality on a long term basis. Providing low-cost but practical and effective solutions to identified problems associated with drinking water supply through applying available scientific and technological inputs. For effective and economic management of water supply systems, the National Water Policy identified thrust areas that need intensifying research efforts. These areas include economical designs for water resource projects, better water management practices, and improvements in operational technology. This study reviews the methods for analysis and design of multi-reservoir multi-junction water transmission networks, which are important components of every water supply system. Recent improvements in design methods that incorporate cost aspects are also discussed in the study.

N Hooda, O Damini (2016) Rural water networks in the developing world are typically branched networks with a single water source. The main design decision for such networks is pipe diameters from a discrete set of commercially available pipe diameters. The larger the pipe diameters, the better the service (pressure), but the higher is the capital cost. In general, each link (the connection between two nodes) can consist of several pipe segments of differing diameters. For such networks, existing design tools solve the constrained-optimization problem heuristically; an ILP formulation is proposed for the special case of one pipe diameter per link. Currently, one can either get an optimal solution for the special case of one piped segment per link or get a non-optimal solution for the general case of multiple pipe segments per link. In this work, we develop a model that solves the general formulation while still maintaining optimality. Our model has an LP formulation. It not only manages to solve the general case optimally; it also has a runtime performance that is better than both the heuristic approach to the general problem as well as the optimal approach to the specific problem (one pipe per link).

2.3.3 Literature Review of Water Distribution Network

To build up information in simulations and analysis of WDN, an in-depth review of literature across this field was examined, which revolves around optimizing the network by understanding the hydraulic parameters. S. Chandramouli et al. (2011) created a modern parameter for assessing, in general, network reliability utilizing fuzzy logic concepts based on the overabundance pressure accessible at demand nodes. The aim was to consolidate the parameter in a two objective optimization model for the design of water distribution network utilizing the combination of EPANET and Genetic Algorithms tool kit within the MATLAB environment. Baiyi Jiang1 et al. (2012) Dynamic pipe network hydraulic modeling realizes the municipal water supply network to optimize effective scheduling and science operations. A point-by-point analysis of the operation utilizing WaterGEMS gives viable illustrations to set up a water supply network based on SCADA data. A. Antonowicz1 et al. (2018) displayed a strategy of utilizing EPANET solver to back oversee water distribution system in Smart City by permitting to perform both single and cyclic simulations. Jiang B et al. (2012) expressed that dynamic pipe network hydraulic modeling is the realization of the municipal water supply network to optimize an effective means of scheduling and operations.

Several hypotheses were carried out, especially in the past two decades. Theofanis P. Lambrou et al. (2014) research presents a low-cost and holistic approach to the water quality monitoring problem for drinking water distribution systems and consumer sites by developing low-cost sensor nodes for real-time and in-pipe monitoring and assessment of water quality. The results demonstrate that this system satisfies the online, in-pipe, low deployment-operation cost, and sound detection accuracy criteria of an ideal early warning system. M. Günthera et al. (2015) introduced an experimental water distribution system (EWDS-TUG) equipped as an intelligent water network comprising measurement and control devices, data collection and communication, data display, and management data fusion and analysis. Stian Bruaset et al. (2017) established a methodology for cohort survival models where historical (empirical) data on decommissioning ages of pipes are used to calibrate survival functions of pipe cohorts according to service level targets and recommended that this process repeated on a 10-year interval in an iterative process where service levels are re-evaluated according to required targets. Herman A. Mahmoud et al. (2018) have discussed and worked on a new pressure- driven analysis PDA method to validate the number of benchmark and real-life networks under different flow conditions, clearly demonstrating its advantages compared to existing methods. Mohd. Abbas H. Abdy Sayyed et al. (2017) illustrated with a benchmark for WDN in this work and funded that a modified GA method converges faster than other penalty application methods. Further, several better solutions are observed to be provided by the proposed methodology provided for the extended network.

Waldo Ojeda-Bustamante et al. (2007) developed a computer tool, Spriter. They transferred it to several WUA of Mexican irrigation districts, which underwater scarcity conditions, estimated irrigation requirements at field scale, and integrated them into several operating levels of the irrigation network.

2.3.4 Literature Review of Water Quality

A literature review revealed that various works had been conducted on the contamination in water distribution networks. According to the literature review given in the first chapter, a detailed account of the corresponding literature in the context of different methods of dealing with contamination in WDNs is given in this chapter.

Santasmasas et al. (2013) designed and installed a decentralized and automatic Micro Bioreactor (MBR) prototype in a facility to treat low-load greywater to be recycled for toilet- flushing applications. Treatment efficiency and to assess the quality Woods et al., (2013) To compare the economic evaluation for water reuse alternatives, an existing decision support system (DSS) was modified to include infrastructure, construction, operation, and maintenance costs for both poTable and reclaimed water at the regional level and the costs associated with the treatment of water to provide both poTable and non-poTable uses of water. Michael E. Try et al. (2002) [9] cited that in contrast to conventional methods that apply disinfectant only at the treatment works, booster disinfection reapplies disinfectant at strategic locations within the distribution system to compensate for the losses as it decays over time. This study addresses locating disinfectant booster stations that minimize the dosage required to maintain residuals throughout the supply network and introduces a valuable parameterization for disinfectant source types. Results obtained suggest that the booster disinfection location model can be applied to realistically scaled problems, but lack of comprehensive application experience precludes any general statements about solution efficiency. Hart William E. et al. (2004) [5] cited that contamination warning systems are a promising approach for mitigating contamination risks in drinking water distribution systems. A critical aspect of a CWS is the strategic placement of online sensors that rapidly detect contaminants. This study reviews the array of optimization-based sensor placement strategies that have been recently proposed. These strategies are critiqued, and several key issues have been identified that need to be addressed in future work. Walter M. Grayman. et al. (2007) [3] cited the ongoing concerns with potential microbial contamination of distribution systems. A quantitative method for assessing alternative microbial sampling programs is described in this study. A distribution system's hydraulic/water quality model is used within a Monte Carlo simulation framework to evaluate the effectiveness of a sampling program. Different paths for entry of the contamination into the distribution system are considered within the probabilistic framework, including cross-connections, pipe breaks, intrusion through transient negative pressures, treatment failure, tank contaminations. The resulting contamination is tracked through the network using EPANET. The method is used to evaluate alternative sampling programs in a hypothetical distribution system. Elin Warn Betanzo et al. (2008) [2] cited that the study presents an analysis of the inactivation provided by disinfectant residuals by using a distribution system model, inactivation, and disinfectant decay models, and conservative assumptions based on available data. Various conditions were modeled, including a range of water quality parameters pH, temperature, inactivation of two microorganisms, and intrusion water dilution ratios. Despite the assumptions inherent in the model, several generalizations were derived from the study. Consequently, systems should consider protection against intrusion when choosing their secondary disinfectant.

Teunis et al. (2010) [14] cited that the study discusses contaminants from the soil surrounding drinking water distribution systems are thought not to enter the drinking water when sufficient internal pressure is maintained. This study presents a risk model for predicting the intrusion and dilution of viruses and their transport to consumers. The virus's random entry and dilution were simulated by embedding the hydraulic model into a Monte Carlo simulation. The probability that a consumer drinks water contaminated with the virus is small, but the virus concentration tends to be high when this happens, and the risk of infection may be considerable. The spatial distribution of infection risk is highly heterogeneous. The presence of chlorine residual reduces the infection risk. R.G. Austin et al. (2010) [1] cited that introducing the traditional risk assessment techniques examines the risk of infection or disease in the context of long-term risk. The current study takes an event-based approach to evaluate the risk of high-impact contamination events. The purpose of this event-based approach is to both evaluate the vulnerability of water distribution systems to such events and to determine the impact that actions taken can have on these events. Risk assessment is done using a Monte Carlo framework, and simulated events are created using EPANET as the hydraulic and water quality model. Walter Grayman.et al. (2012) [4] cited that one of the most common uses of water distribution systems and water quality modeling is to simulate the chlorine residual in the network. EPANET is used to model the decay of chlorine using a bulk decay rate and a wall decay coefficient. In EPANET, the bulk decay rate is defined as a single constant for the entire distribution system. However, if multiple sources within a distribution system produce finished water with different chlorine decay characteristics, it is not possible within EPANET to accurately represent the decay in areas where the water from multiple sources is blended. The multiple species extension (MSX) to EPANET allows the user to create a simple text file representing the bulk decay. Rafi Schwartz. et al. (2013) [12] cited that event detection is currently one of the most challenging water distribution systems analyses. The problem is how regular online hydraulic and water quality measurements can be efficiently utilized to designate a contamination occurrence. To date, all experiments on testing event detection models are based on generating arbitrary manipulations on water quality parameters. In this study, EPANET-MSX was employed for simulating complex contamination intrusions for enhancing event detection model verification. The objective is to simulate contamination events at locations in which event detection instrumentations are placed.

L. Monteiro. et al. (2013) [8] cited that the modeling of chlorine residual in water supply systems is of great importance in managing disinfectant concentrations throughout the network. First-order decay kinetics is currently often used to describe both bulk and wall chlorine consumption. However, more complex approaches have been proposed, namely, the 2R model that has been reported as yielding better accuracy for simulating chlorine bulk decay in laboratory tests. In the current study, the performance of the 2R model and first and nth order decay kinetics was assessed for full-scale modeling of chlorine in a transmission system. Results have shown that a similar level of accuracy can be achieved with the three tested kinetic models, provided that a good calibration of the wall decay coefficient is accomplished. M. Tamer Ayvaz. et al. (2014) [11] cited that the current literature includes several studies for optimal scheduling and locating booster chlorination stations. In this study, a new simulation- optimization approach is proposed for booster chlorination network design. A hybrid

GA-LP (genetic algorithm-linear programming) model is proposed, determining the optimum number of booster stations, locations, and injection dosages while maintaining residual chlorine concentrations within desired limits throughout the WDN. Weidong Zhang. et al. (2015) [15] cited that this study presents a new mathematical model of bacterial regrowth in distribution systems. The microbial process description has been simplified from previous models based on a sensitivity analysis. The alternating split-operator algorithm is used to solve this model. The use of the model to understand the interaction among critical parameters affecting bacterial regrowth is illustrated for a simple hypothetical network. R V Goyal. et al. (2016) [13] cited that the optimum location of booster chlorination stations is worked out along with the scheduling of booster chlorination stations for the steady-state flow conditions for the drinking water distribution system (DWDS). Linear programming in Excel is utilized to formulate the model by coupling the simulation results of EPANET software for residual chlorine at critical nodes and optimal location of booster chlorination stations, using different combinations of booster stations. The optimal location of booster chlorination stations, combined with optimal scheduling of booster stations using optimization method, can serve as an essential decision- making tool for any water supply authority for selecting the appropriate location, several booster stations, application of optimum chlorine mass rate, and overall cost reduction. N. Khanal. et al. (2016) [7] cited that the drinking water distribution systems are vulnerable to various terrorist threats, including physical attack, cyber disruption, and biochemical contamination. Efficacy of the attack logistics: injection location, injection duration, and injection timing were examined in the entire model of the Cherry Hills – Brushy Plain water distribution network. The effect of tank mixing on nodal mass loading and population exposure is studied. The results generated are of statistical significance and are easily understandable by utilities and designers. Muhammad Nadeem Sharif. et al. (2016) [10] cited that the water quality management in distribution networks is directly related to the spatial distribution of chlorine boosters and their dosages. Water chlorination is essential to reduce the effects of bacterial and other microbiological contaminants, but a higher dosage of chlorine generates harmful by-products in addition to changes in drinking water's taste and odor. Hence optimization of chlorine dosage is necessary to decrease the microbial contaminants that affect water quality. Once the chlorine threshold is determined for a microbial contaminant, it would % help decision-makers suggest optimal values.

2.3.5 Security Management

Theofanis P. Lambrou et al. (2014) research presents a low-cost and holistic approach to the water quality monitoring problem for drinking water distribution systems and consumer sites by developing low-cost sensor nodes for real-time and in-pipe monitoring and assessment of water quality. The results demonstrate that this system satisfies the online, in-pipe, low deployment-operation cost, and sound detection accuracy criteria of an ideal early warning system. M. Günthera et al. (2015) introduced an experimental water distribution system (EWDS-TUG) equipped as an intelligent water network comprising measurement and control devices, data collection and communication, data display, and management data fusion and analysis.

Stian Bruaset et al.(2017) established a methodology for cohort survival models where historical (empirical) data on decommissioning ages of pipes are used to calibrate survival functions of pipe cohorts according to service level targets and recommended that this process be repeated on a 10-year interval in an iterative process where service levels are re-evaluated according to required targets. Mohd. Abbas H. Abdy Sayyed et al. (2017) illustrated with a benchmark for WDN in this work and funded that a modified GA method converges faster than other penalty application methods. Further, several better solutions are observed to be provided by the proposed methodology provided for the extended network.

2.4 MOBILE APPLICATION

2.4.1 Overview

The following section presents a summary of the work carried out by various researchers on the analysis and optimization of data collection applications. Much research has already been carried out in research data collection, providing a greater depth into vast and complex study material.

2.4.2 Summary of Literature Review Conducted

Frances Jeffrey-Coker et al. (2010), describe the phones were equipped with ODK software, which immediately digitizes data for analysis, allows for remote monitoring of collection process, facilitates data gathering, and avoids study surveys reducing survey time during the study. Carl Hartung et al. (2012), cited an open-source suite of tools designed to build information services for developing regions. It currently provides four tools to this end collect, aggregate, voice, and build. They describe four deployments and demonstrate how a decision made in the system architecture of ODK enables services that can both push and pull information in developing regions. MA Brovelli et al. (2015), describes that the study presented aims to visualize volunteered data in 3D, considering aspects of the general public using free and open-source software. The data collected through various volunteered geographic information (VGI) might be in a different format. The project developed aims to visualize and query data collected through the ODK platform and a cross-platform application. Evan Newcomer et al. (2017), the feasibility of reducing greywater for domestic activities in a rural area was evaluated from a microbiological and public acceptance perspective. They surveyed 128 households and collected 28 domestic greywater samples to check the properties as per WHO guidelines. MA Brovelli et al. (2017) cited that using the ODK application, the data is collected, and NASA makes the open-source framework NASA World Wind. It mainly comes in two versions the java version and the web version. Ming-Gong Lee et al. (2016), cited a new interface to collect field data, including text, numeric, GPS, photos, and video, to introduce the plant classification in the city park named 18 peak mountains is collected. The finding and research gaps obtained in the literature over the past few decades by various researches show that a well-planned network can be appropriately optimized if the layout is designed keeping factors such as sustainability, optimization, predicting water leakage, aging of pipes—the holistic approach along with applying various algorithms for simulations and modeling.

CHAPTER - 03

Methodology

3.1 DATA COLLECTION AND SYSTEM REQUIREMENTS

In this research, a few data is required to be downloaded from official websites; hence, it is necessary to understand the organization, system, and procedure. Therefore, there are three significant discussions done below, and they are:

a. Bhuvan Portal – Indian Space Research Organisation (ISRO)

b. Toposheet – Survey of India

c. LISS-III Satellite Imagery

3.1.1 Bhuvan Portal – Indian Space Research Organisation (ISRO)

Indian Space Research Organization (ISRO) has launched a web-based utility platform called BHUVAN (Source: https://bhuvan.nrsc.gov.in/bhuvan_links.php#), by which the students, researchers, scientists can access, explore and download many contents of map sets freely for research and scientific purpose. The content is offered in four regional languages, i.e., English, Hindi, Telugu and Tamil.

Various thematic maps can be explored in agriculture, disasters, water resources, land use land cover restricted to Indian boundaries. Various sections of the Government of India (GoI) are associated with Bhuvan to use this platform in Geospatial Technology, where the data is available as layers of information for visualization for public consumption. These layers include geospatial information census data, island information, school Bhuvan, toll information, groundwater, pipeline grid, watersheds, urban survey of all states and union territories of India.

3.2 TOPOSHEET – SURVEY OF INDIA (SOI)

Survey of India is one of the earliest scientific departments set up in 1767 under the Government of India (GoI) – Department of Science and Technology (DST), which deals with mapping organizations and the National Survey.

3.2.1 Toposheet Identification Procedure

Toposheet is a pictorial representation of a graphical portrayal of various objects on the earth's surface. For various engineering works, these maps give preliminary information of the terrain and are hence valuable. Survey of India has provided a specific numbering system for these topographic maps to identify a particular area that falls under the interest, and below mentioned is the technique to select it:

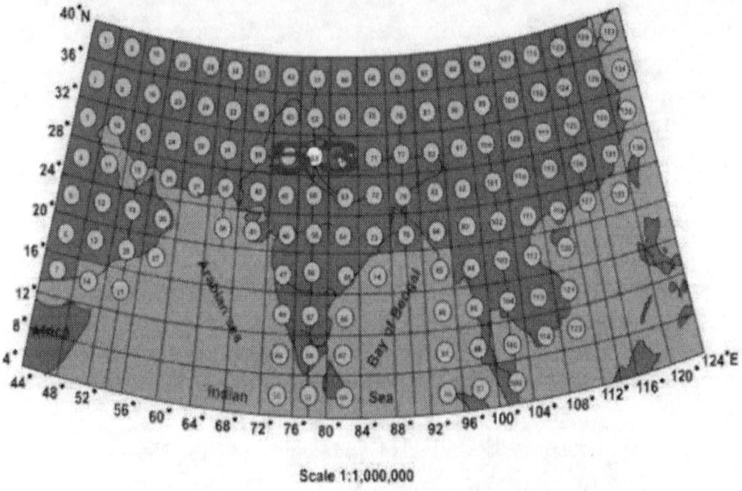

Figure 3.1: Index to the International Series for Base Map
(**Source:** http://gyaninformationpedia.blogspot.com)

1. An international series of (4 to 40)° North Latitude and (44 to 124)° East Longitude is considered as a base map with a scale of 1: 1,000,0000. These base maps are further divided as sections of 4° (latitude) x 4° (longitude) and nominated from extreme N-W as 1 to 136 as shown in Figure 3.1, in which only land areas are covered, and any 4° which lies undersea is not considered.

2. Topographic maps which fall under the Indian subcontinent are divided further into 16 sections (4 x 4) rows and columns each representing (1° x 1°) latitude x longitude with a scale of 1:250,000 scale with the alphabet "A" noted at the extreme N-W corner and ending with "P" at extreme S-E corner written column-wise as shown in Figure 3.2.
3. Further, each sheet is subdivided into four parts (2x2) into rows and columns with (30'x30') latitude x longitude scaling 1:100,000 representing NW, NE, SW, and SE as cardinal directions and identified as 53M/SE as shown in Figure 3.2.
4. Further, each sheet is subdivided into 16 sheets (4x4) in rows and columns with (15'x15') latitude x longitude scaling 1:50,000 with number "1" noted at the extreme N- W corner and ending with "16" at extreme S-E corner written column-wise and represented as 53 b/3 as shown in Figure 3.2.
5. Each sheet is divided into four parts (2x2) in rows and columns with (7'1/2 x 7'1/2) latitude x longitude scaling 1:50,000 representing NW, NE, SW, and SE as cardinal directions and identified as 53 O/14//NE as shown in Figure 3.2.
6. As mentioned above, the toposheet can be upgraded and updated as per the project requirement and can be acquired at the scale for most of the areas in India, and further, if required, surveying can be done more if large-scale maps are needed for special projects.

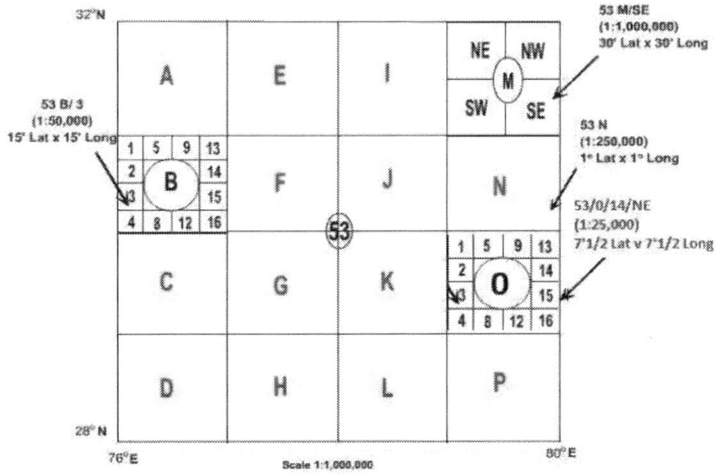

Figure 3.2: Map Numbering for Indian Topographic numbering system
(**Source:** http://gyaninformationpedia.blogspot.com)

3.3 LISS-III

Linear Imaging Self Scanning Sensor-III (LISS-III) is a camera that has four spectral bands, three of which are invisible & near-infrared regions and one in the shortwave near-infrared reign of the electromagnetic spectrum. LISS_III camera in shortwave near-infrared band lies in between 1.55 to 1.7 microns with 23.5 spatial resolution. European Petroleum Survey Group (EPSG) provides accurate information documents related to map datums and projections in a database. EPSG 3857 coordinate system is used, which refers to latitude and longitude on WGS reference ellipsoid.

LISS-III has various specifications such as spatial resolution, spectral bands (4 bands) within the electromagnetic spectrum, swath width (km), power consumptions (Watts). Table 3.1 below all the specifications of the LISS-III.

Table 3.1: Summary of the Indian Remote Sensing (IRS-P6) instrument parameters

Parameter of Satellite Image	LISS-3 Image Data Specifications
Instantaneous Field of View (IFV)	23.5 m -Spatial resolution
Spectral bands (μm)	B2: 0.52-0.59, (green), B3: 0.62-0.68, (red), B4: 0.77-0.86, (NIR), B5: 1.55-1.70 (SWIR)
Swath width	141 km
Detector line arrays x No of elements	4 x 6,000
Data quantization	7 bit (VNIR), 10 bit (SWIR)
Square wave response at Nyquist	B2> 0.40, B3>, B4> 0.35, B5> 0.20
Power consumption	70 W
Instrument mass	106.1 kg
Data rate	52.5Mbit /s

(**Source**: https://earth.esa.int/web/eoportal/satellite-missions/i/irs-p6)

3.4 SPATIAL DATA

For collecting spatial data, a satellite image of the said location was downloaded from the Bhuvan website, which is an Indian geo platform of ISRO (Indian Space Research Organization. From open data archive, NRSC/ISRO Open data and product archive facilitates the user selecting, browsing, and downloading data from portal under three major categories.

1. Satellite / Sensor
2. Theme/product and
3. Program /project.

3.4.1 Procedure for Downloading Cartosat-I: DEM Satellite Data

As mentioned above, out of the various satellite data, Cartosat 1 and LISS III satellite data is used for this research study, mainly showing PAN images of 2.5 km resolution and three spectral bands with the 23.5-meter resolution.

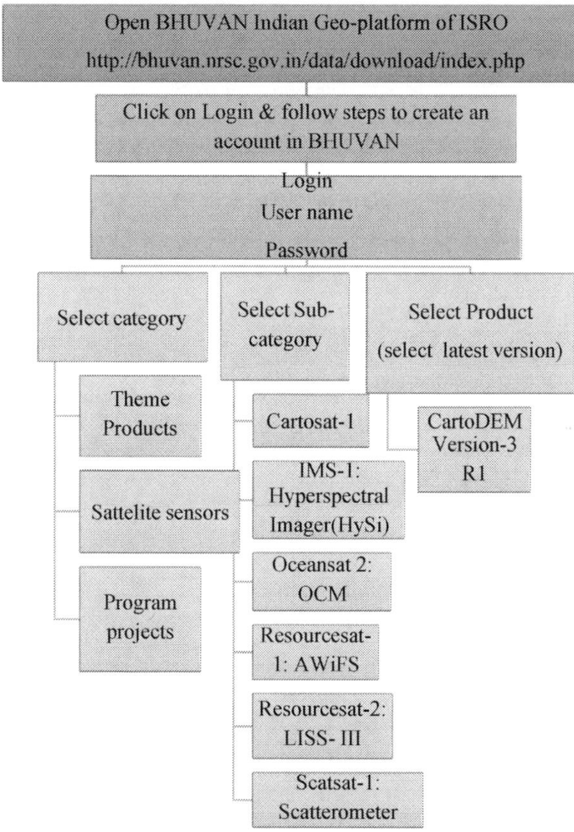

Figure 3.3: Methodology to download the satellite data

Both the satellite data are downloaded through the national geo platform of ISRO: BHUVAN. Figure 3.3 shows the methodology to download the satellite data, and the technique of selecting the satellite

/ Sensor category and subcategory by drop-down menu. After opening the Bhuvan portal through (http://bhuvan.nrsc.gov.in/) their official website, we need to log in by creating an account with an authenticated email id which would have its unique user name and password.

As we proceed ahead, the National Remote Sensing Centre (NRSC) /ISRO Open data and product archive facilitates the user selecting, browsing, and downloading data from this portal under three major categories. (Source: http://bhuvan.nrsc.gov.in/data/download/index.php)

a) Satellite / Sensor
b) Theme/product and
c) Program /project.

Under the subcategory, we have to select the Satellite/Sensor, which would suit our research purpose.

a) Cartosat-1*
b) IMS-1: Hyperspectral Imager (HySi)
c) Oceansat 2: OCM
d) Resourcesat-1: AWiFS
e) Resourcesat-2: LISS- III*
f) Scatsat-1: Scatter meter

Note: * In this research study, Cartosat-1 and Resourcesat-1: LISS_III satellite data is utilized to drape False Color Composite (FCC) image over Digital Elevation Model (DEM) in QGIS an Open-source GIS software.

3.4.2 Cartosat-1

Cartosat-1 or IRS-P5 is a stereoscopic Earth observation satellite in a sun-synchronous orbit and the first one of the Cartosat series of satellites. Cartosat-1 carries two state-of-the-art panchromatic (PAN) cameras that take black and white stereoscopic pictures of the earth in the visible region of the electromagnetic spectrum. The swath covered by these high-resolution PAN cameras is 30 km, and their spatial resolution is 2.5 meters. The payload is designated to perform cadastral mapping, cartography, terrain modeling, and other applications. The Cartosat-1 Digital Elevation Model (Carto DEM) is a National DEM developed by the Indian Space Research Organization (ISRO). It is derived from the Cartosat-1 stereo payload launched in May 2005. Four methods can download the satellite data, which is in image form. Figure 3.9 shows the Bounding box method was used where the study's latitude and longitude are entered to get the data.

Table3.2 shows Satellite parameters and specifications details of Resources-I.

Table 3.2: Satellite parameters and specification details of Cartosat -I .

Parameter of Satellite Image	Image Data Specifications
Date of Launch	5th May 2005
Launching Site	Satish Dhawan Space Centre in Sriharikota
Launching Vehicle	Polar Satellite Launch Vehicle (C6)
Orbit details	Polar Sun Synchronous-618 km
Orbital Period	97 minutes
No. of orbit/day	14 orbits
Equator local crossing time	10:30 AM
Receptivity (in days)	126 days
Revisit (in days)	Five days
Mission life (in years)	Five years

(**Source:** https://www.isro.gov.in/Spacecraft/cartosat-1)

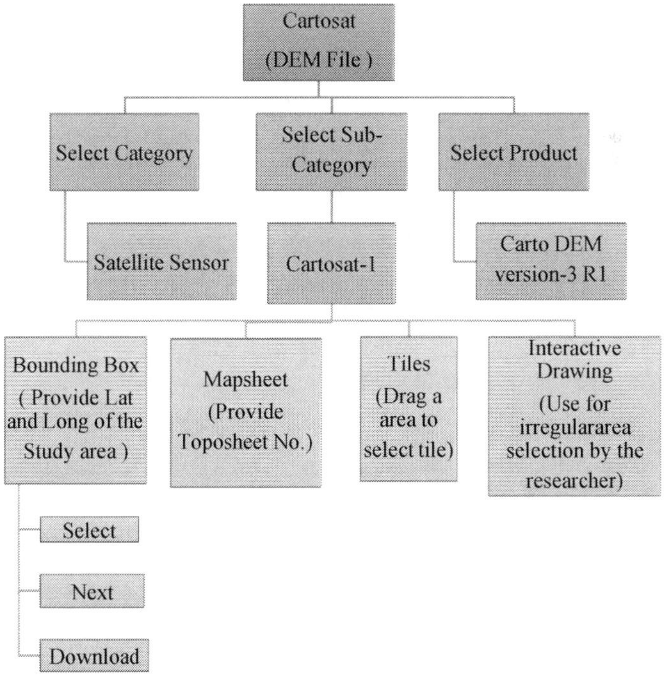

Figure 3.4: Methodology to download the Cartosat I Satellite data

The user interface box is shown in Figure 3.4, which shows four methods bounding box, Map sheets, tiles, and interactive drawing methods are downloaded at the user's convenience. In the Bounding box method, the minimum and maximum latitude and longitude of the area of interest must be fed. In the map sheet, the sheet number, if known, has to be fed, whereas in tiles directly, the square tile needs to be selected on India's map, and in Interactive drawing with the use of editing tool provided the boundaries of the research area needs to be drawn.

3.4.3 Procedure for downloading Resourcesat-I: LISS_III Satellite Data

Table 3.3: Resourcesat-1 Satellite details

Parameter of Satellite Image	Image Data Specifications
Date of Launch	17th October 2003
Launching Site	Satish Dhawan Space Centre in Sriharikota
Launching Vehicle	Polar Satellite Launch Vehicle (C5)
Orbit details	Polar Sun Synchronous
Orbital Period	101.35 minutes
No. of orbit/day	14 orbits
Equator local crossing time	10:30 AM
Receptivity (in days)	126 days
Revisit (in days)	Five days
Mission life (in years)	Five years

(**Source:** https://www.isro.gov.in/Spacecraft/cartosat-1)

3.4.4 Resourcesat 1 - LISS III

The LISS-III is a multi-spectral camera operating in four spectral bands, three in the visible and near-infrared and one in the SWIR region case of IRS- 1C/1D. The new feature in the IRS-P6 LISS-III camera is the SWIR band (1.55 to 1.7 microns), which provides data with a spatial resolution of 23.5 m, unlike in IRS-1C/1D (where the spatial resolution is 70.5 m) and a swath of 141 km.

Figure 3.5 shows a detailed procedure to download the Resourcesat I: LISS III satellite data. Resourcesat-1(Satellite) and LISS-3 (Sensor) was considered for this study as it merges with the objective of the research.

Table 3.3 below gives a brief description of the tile, projection, and image-related spatial data. This data is received from Linear Imaging and Self Scanning Sensor (LISS), which operates in three spectral bands in VNIR and one band in SWIR with 24- meter spatial resolution and a swath of 141 km. Table 3.3 gives satellite parameters and specification details of Resourcesat-I tile-related data of the Resourcesat-1(LISS-III) sensor a tile size is 15' x 15' with a Collar of 40 pixels per the Survey of India (SoI) Open Series Map (OSM) conventional pattern. The data projection used in QGIS is by latitude longitude commonly termed as Geographic coordinate system with Datum as World Geodetic System (WGS84) with a resolution of 0.000225 degrees. The coordinate system, which is projected from the sphere or ellipsoid surface to a flat surface, is termed EPSG:3857 (European Petroleum Survey Group) and is used by all tile providers commercially for free.

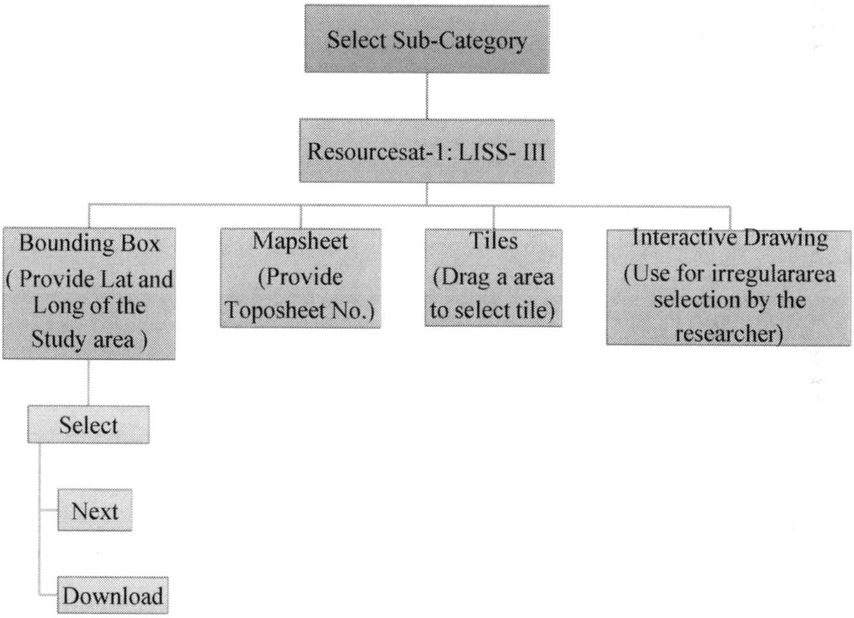

Figure 3.5: Methodology to download Resourcesat I: LISS III Satellite data

The image file format is of Tagged Imaged Format Fill (TIFF image), commonly used to exchange bitmap (raster graphics) images between application programs, commonly scanner images. These TIFF files have four bands, as mentioned before, namely bands 2,3,4, and 5 with 8 bits radiometric resolution. Name of the dataset used is L3_SAT_8B_v1_73E18.75N_E43H01 _01feb13 with tile name E43H01

3.5 STUDY AREA DATASETS

This research accentuation the application of Remote sensing and GIS and its impedances with NDVI and EVI.

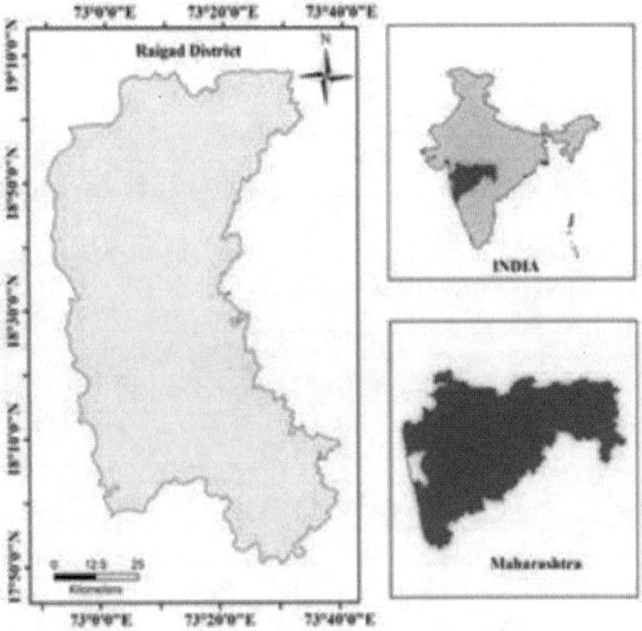

Figure 3.6: Study Area – Raigad, Maharashtra (India)

The region of interest is chosen considering the geographic and environmental perspectives. The study area is near Navi Mumbai, Maharashtra, India, as Figure 3. 6. It lies in between 18°.58'.44.61" to 19°.0'.57.16" latitude and 73°.02'.54" to 73°.05'.39.61" longitude in Maharashtra, India on a development site of rehabilitation. Refer to Figure 4 and 5, which shows the study area captured in 1998 and 2009 by Landsat 5 and 2020 by Landsat 8, respectively, as shown in Figure 3.7. Due to the rapid growth of Remote Sensing, GIS, and other geospatial technologies in various fields, the interdependence and the relationship between man and the environment, which is vital in today's artificial world, are helping in making important decisions for the growth of humankind. This research primarily observes the fluctuations in vegetation growth over the past two decades and has helped change detection in the study area.

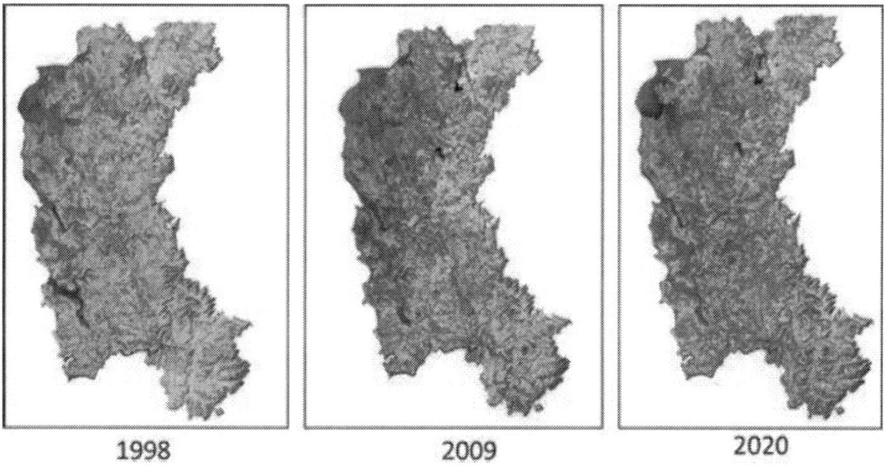

Figure 3.7: Decadal Satellite Image of Raigad, Maharashtra (India) - Landsat 5 and 8

3.6 WATER SUPPLY TO CIDCO AREA

3.6.1 Water Supply Various Nodes

The various departments such as Patalganga, Morbe, Hetwane, and Barvi have been sanctioned some water quota in MLD, and the actual availability at the nodes is relatively more than required. For more details, refer to Table 3.4. It can be seen that the overall sanctioned quota by CIDCO is about 213 MLD, and the actual availability at the endpoint is 190 MLD, i.e., about 23 MLD less the required.

Table 3.4: Water Supply at various Nodes

Sr. No.	Source / Department	Sanctioned quota to CIDCO	Actually, available at the endpoint
1	Patalganga / MJP	85 MLD	69 MLD
2	Morbe / NMMC	35 MLD	4 MLD
3	Hetwane / CIDCO	100 MLD	86 MLD
4	Barvi / MIDC	3 MLD	1 MLD
	Total	213 MLD	190 MLD

(**Source:** CIDCO)

3.6.2 Water demand in CIDCO areas

The total CIDCO area is divided into 12 zones Dronagiri, Ulwe, Kharghar, and Kharghar hills, Navase – Phase I / II, Kamothe,

Karanjade, Kalundre, Taloja, Kalamboli, Panvel, Airport, Pushpak 2.5 FSI, respectively. As per data, the total ultimate water demand is 700 MLD at the start of the project, and an increase of 210 MLD is added in the year 2014. However, the water supply consumption in 2014 is about 190 MLD which clearly shown a deficit of 20 MLD, as shown in Table 3.5.

3.6.3 Immediate Steps to Meet the Water Demand Scenario in 2014

To find a solution to the deficit of 20 MLD, a sanction of an additional 100 MLD from Konkan Irrigation Development Corporation (KIDC) and 50 MLD as the quota from Hetwane is added so as 150 MLD compensates the total deficit.

Table 3.5: Water demand in 2014

Sr. No.	Node	Ultimate water demand MLD	Water demand in 2014 in MLD	Water supply consumption 2014 in MLD	Water supply Deficit in MLD
1	Dronagiri	100	12	10	2
2	Ulwe	80	7	6	1
3	Kharghar & Kharghar hills	130	70	66	4
4	Navase - Ph-I / II	25	1	0.5	0.5
5	Kamothe	95	40	35	5
6	Karanjade	35	1	-	1
7	Kalundre	5	1	0.5	0.5
8	Talajo	40	4	3	1
9	Kalamboli	40	30	27	2
10	Panvel	60	44	42	3
11	Airport	40	-	-	-
12	Pushpak 2.5 FSI	50	-	-	-
	Total	700	210	190	20

(**Source:** CIDCO)

The existing pumping of raw water at WTP Jite needs to be updated to increase an additional 150 MLD from the Hetwane dam. Water supply to enroute villages from Hetwane dam to PCB of 25 MLD to explore extra quota from KIDC irrigation department.

3.6.4 Water Procurement Rates

The water is available from various sources such as Patalganga MJP, Morbe NMMC, MIDC, and Hetwane CIDCO respectively have different water procurement rates as shown in Table 3.6.

Table 3.6: Water Procurement rates

Sr. No.	Source / Department	Water Procurement rate per cum.
1	MJP	8.00
2	NMMC	15.00
3	MIDC	16.25
4	CIDCO	6.75

(**Source:** CIDCO)

3.6.5 CIDCO Tariff Rates for Consumers Since 2000

The billing process for water rates to various consumers from 2006, i.e., CIDCO tariff water rates for metered consumers, is shown in Table 3.7 and Domestic rates for flat rate (CIDCO housing) consumers are shown in the Table 3.8 respectively.

Table 3.7: CIDCO tariff water rates for metered consumers

Sr. No.	Types of Consumers	Water rate/CUM in Rs.
1	Commercial / Construction	35.00
2	Social / Institution / ChariTable	14.00
3	Domestic rate	
a	up to 20 CUM / month / family	4.75
b	up to 27 CUM / month / family	6.00
c	exceeding 27 CUM / month / family	7.00

(**Source:** CIDCO)

3.6.6 Details of Hetwane Water Supply Scheme (Source: CIDCO)

1. CIDCO tunnel intake structure and allied works to draw water from Hetwane dam.
2. Provision of raw water gravity pipeline from Hetwane dame to Jite.

3. Raw water pump, pump house, and allied works at the site to pump water 150 MLD.
4. WTP with pure water sump and pump house at the site at 150 MLD capacity.
5. Pure water rising main at 10.40 km from WTP to Digha.
6. Brake pressure tank at Dighati: 750 cum.

Table 3.8: Domestic rates for flat rate (CIDCO housing) consumers

Sr. No.	Types of Consumers	Water rate/CUM in (Rs./ Month)
1	up to 25 CUM	75.00
2	25-40 CUM	82.00
3	40-50 CUM	99.00
4	50-60 CUM	120.00
5	60-70 CUM	125.00
6	70-80 CUM	125.00

(**Source:** CIDCO)

3.7 SYSTEM REQUIREMENTS

The data collection involves the following details

- Hardware and Software requirements
- Plug-in requirements
- Other requirements

3.7.1 Hardware and Software Requirements

Before starting the process of georeferencing, a few hardware and software are required, and, i.e., a Personal Computer installed with Windows XP/Vista operating system with a Central Processing Unit (CPU) of 4.24GHz+ (plus) or Advanced Micro Devices (AMD) of 2400xp+ (plus). A minimum of 512 MB Random Access Memory (RAM) and a hard disc of a minimum of 2GB free space is required, along with a Network speed of 768 Kbits/sec. A 3D capable Graphic card with 32MB of VRAM is sufficient, and a screen of 128 x 1024, 32-bit actual color would be required.

3.7.2 Plug-In Requirements

For browsing the Bhuvan website, the Bhuvan plug-in (with administrative privileges only) is required to be downloaded from this website, and after registration, DirectX 9.0 or upgraded version (www.microsoft.com/windows/directx/) along with MS.NET framework 2.0 or higher for plug-in installation would be required.

3.7.3 Other Requirements:

Authenticated email account for downloading satellite data from Bhuvan is required, along with QGIS software (latest version) can be downloaded from the internet as it is available on an open-source platform along with Firefox or Chrome web browser.

3.8 METHODOLOGY

Figure 3.8 shows the overall research methodology.

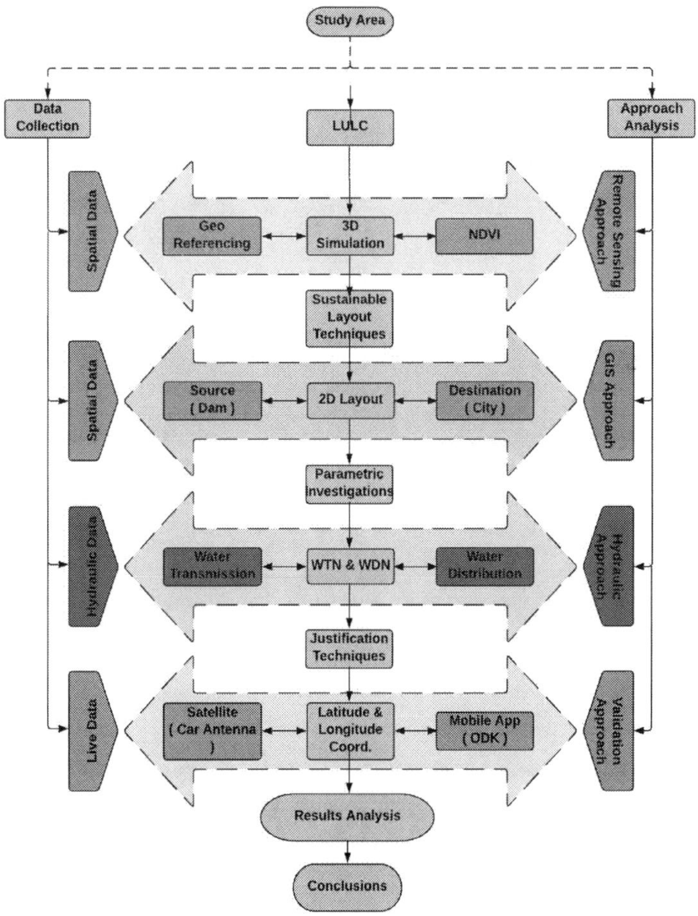

Figure 3.8: Overall Research Methodology

The whole methodology in a thesis research is classified into four major divisions, which are supported by respective data collection techniques and approach analysis Land Use Land Cover, sustainable layout techniques, parametric investigations and justification techniques The above flowchart, as shown in Figure 3.8, explains the four divisions mentioned above.

3.8.1. Land Use Land Cover (LULC)

The area of interest is near the upcoming smart city near Mumbai, and due to rapid changes in urbanizations in the past two decades, change detection analysis was required to be implemented.

Temporal satellite data is used in this research, which stores time, time interval, and other vital information in the datasets in character string formats, improving its consistency and validation requirements. Data in the hour, min, sec, and micro sec are stored from GMT in time zone offset format. The methodology of this research. Refer to Figure 3.9.

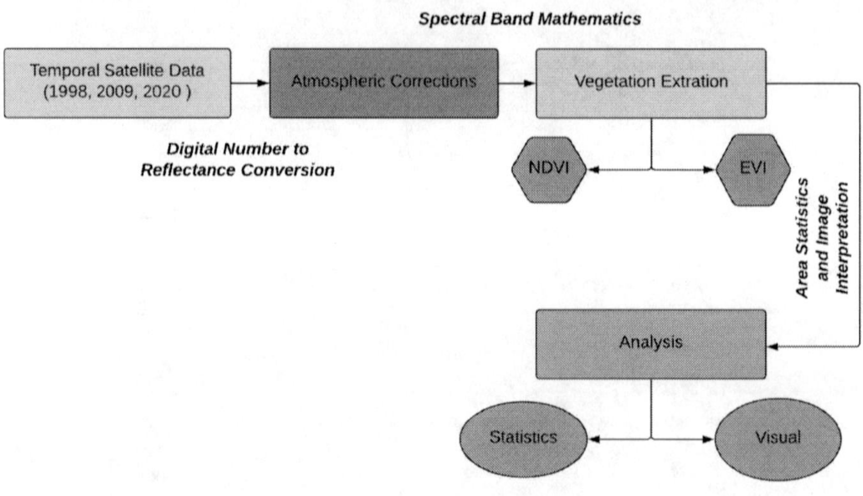

Figure 3.9: LULC Methodology

This research has taken temporal satellite data of the study area by Landsat 5 and Landsat 8 acquired at different times from 1998 to 2020 with 11 years intervals, i.e., Landsat 5(1998), Landsat 5 (2009), and Landsat 8 (2020), respectively. The sensors such as Enhanced Thematic Mapper Plus (ETM+) and Landsat Thematic Mapper (TN) convert the data captured from reflected solar energy to radiance.

These radiances are then reframed or rescaled to digital number (DN) in 8-bit, i.e., 28, and vary from 0-255.

These temporal satellite data sets are then processed for atmospheric correction wherein the reflectance value of the imagery taken by the Landsat sensor is observed, and the effects due to atmosphere are corrected. In optical remote sensing, the atmospheric effects are vital and complex as it alters its spectral nature of radiation which reaches the sensor. The atmospherically corrected data is later processed for Spectral band mathematics. The computer that displays the enhancements of images can be altered as these images are impermanent and usually saved in software for project documents. Statistical and mathematical methods are applied to achieve spectral relationships between bands and produce new data in raster products archivable to interpret vegetation index. Normalized Difference Vegetation Index (NDVI) and Enhanced Vegetative Index (EVI) are a few standard extraction methods used in vegetation analysis for time-series datasets.

3.8.2. Normalized Difference Vegetation Index (NDVI)

Calculating NDVI is simple and easy to understand. It is the ratio of subtraction in the numerator to addition in denominator between near-infrared (NIR) and red bands, respectively, as shown in equation 3.1 , and it ranges from (-1) to (+1). Usually, a typical green vegetation NDVI ranges from a minimum of 0.2 to a maximum of 0.9. Figure 3.10 shows NDVI for 1998, 2009, and 2020 resp.

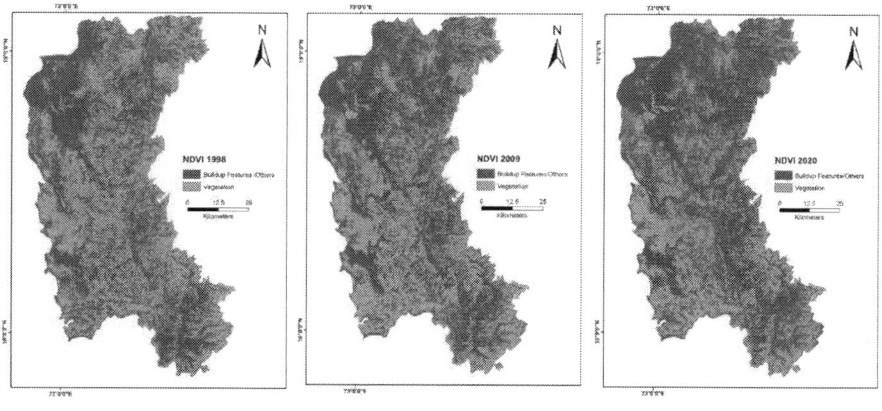

1998 2009 2020

Figure 3.10: Normalized Difference Vegetation Index (NDVI) Dynamics

$$\mathbf{N.D.V.I} = \frac{\mathbf{IR - R}}{\mathbf{IR + R}} \quad ...(3.1)$$

Where; IR and R are Infra-Red band and Red band, respectively

NDVI is applied in this research to identify various vegetation in agricultural fields, forests, gardens, open green grounds and other vegetative growth phases. Forests and rich green vegetation have shown much brighter ridgeline and high NDVI values, whereas uncultivated fields show low NDVI values and are usually dark. NDVI highlights impervious surfaces and water bodies in the dark but cannot differentiate between the two based on values alone. However, the interpretation of elements by shapes can be possible through a visual approach, such as classifying between a meandering river or a road that is going straight.

3.8.3 Enhanced Vegetative Index (EVI)

Enhanced Vegetative Index (EVI) is similar to NDVI. It can be referred to as an optimized vegetative index that helps identify the vegetative signals with advanced sensitivity in regions where biomass is high and enhanced monitoring of vegetation by de-coupling signals of canopy background and reduction cases due to influences on the atmosphere. The formulae to calculate EVI are shown in equation 3.2. Figure 3.11 shows EVI for 1998, 2009, and 2020 years resp.

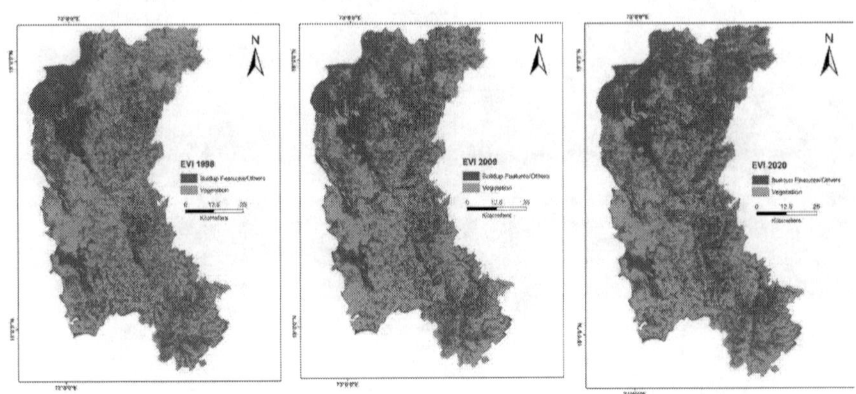

1998 **2009** **2020**

Figure 3.11: Enhanced Vegetation Index (EVI) Dynamics

$$\mathbf{E.V.I} = \mathbf{G\ x} \frac{\mathbf{(NIR - R\,)}}{\mathbf{(NIR + (C1\ x\ R\,) - (C2\ x\ B) + L\,)}} \quad ...(3.2)$$

Where;

E.V.I is Enhanced Vegetative Index.

NIR and IR are Near Infra-Red band and Red band, respectively.

B and L indicate the blue band and Canopy background value, respectively. C1 and C2 are atmospheric correction coefficients.

EVI combines canopy background (L value) and coefficients which are used for correcting atmospheric resistance (C values) along with values from the blue band (B). By improving the output, the saturation, atmospheric noise, and background noise are reduced to a greater extent to enhance the index calculation between R and NIR.

In Landsat 4 to 7,

$$\mathbf{E.V.I.} = \mathbf{2.5\,x} \frac{(\mathbf{Band\ 4} - \mathbf{Band\ 3})}{(\mathbf{Band\ 4} + (\mathbf{6\,x\,Band\ 3}) - (\mathbf{7.5\,x\,Band\ 1}) + \mathbf{1})} \quad \text{...(3.3)}$$

Where

Band 1 : Visible (0.45 – 0.52 μm) 30m.

Band 2 : Visible (0.52 – 0.60 μm) 30m.

Band 3 : Visible (0.63 – 0.69 μm) 30m.

Band 4 : Near Infra-Red (0.77 – 0.90 μm) 30m.

Band 5 : Short Wave Infra-Red (1.55 – 1.75 μm) 30m.

Band 6 : Thermal (10.40 – 12.50 μm) 60m : Low gain / High gain

In Landsat 8,

$$\mathbf{E.V.I.} = \mathbf{2.5\,x} \frac{(\mathbf{Band\ 5} - \mathbf{Band\ 4})}{(\mathbf{Band\ 5} + (\mathbf{6\,x\,Band\ 4}) - (\mathbf{7.5\,x\,Band\ 2}) + \mathbf{1})} \quad \text{...(3.4)}$$

Where

Band 1 : Coastal/Aerosol (0.433 – 0.453 μm) 30m.

Band 2 : Visible Blue (0.450 – 0.515 μm) 30m.

Band 3 : Visible Green (0.525 – 0.600 μm) 30m.

Band 4 : Visible Red (0.630 – 0.680 μm) 30m.

Band 5 : Visible Infra Red (0.845 – 0.855 μm) 30m.

Band 6 : Short wave Infra Red (1.56 – 1.66 μm) 30m.

Results vary the distinct classification of the land Use and Land Cover across the boundary of the study area. Hence this research provides the latest LULC map. This would help in suggesting the best possible route by which the WTN can be laid in those lands where its infrastructural requirements, technical planning as well as designing aspect can be considered at their best without disturbing the current scenario keeping in view the environmental as social aspects.

In the future, these LULC maps can be used to detect leakages, water theft, and the branching of WTN lines, wherein these maps would act as the master plans for various other future infrastructural projects.

3.8.4 Normalized Difference Vegetation Index (NDVI) and Enhanced Vegetative Index (EVI) Dynamics

NDVI (Figure 3.12) and EVI (Figure 3.13) shows the change detection analysis, and the comparison of the study area (between years 1998 -2009-2020) shows a drastic growth in urbanization.

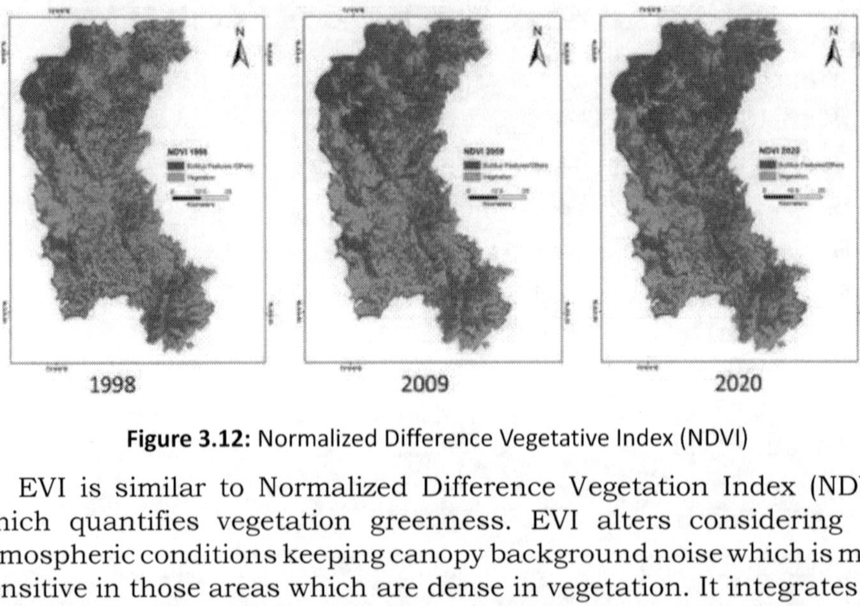

Figure 3.12: Normalized Difference Vegetative Index (NDVI)

EVI is similar to Normalized Difference Vegetation Index (NDVI), which quantifies vegetation greenness. EVI alters considering few atmospheric conditions keeping canopy background noise which is more sensitive in those areas which are dense in vegetation. It integrates "L" and "C" values which are applied for canopy background adjustment and atmospheric resistance coefficients, respectively, along with blue band (B) values. While decreasing saturation, atmospheric, and background noise in most cases, these enhancements help determine the index calculation, which is a ratio between Red and Near Infra-Red values.

According to this research, the change detection analysis has shown decreased vegetation index in the past two decades, from 1998 till 2020. Urban sprawl has increased from 18.5 % to 52.1 %, indicating a decline in vegetation from 88.5 % to 47.9 %.

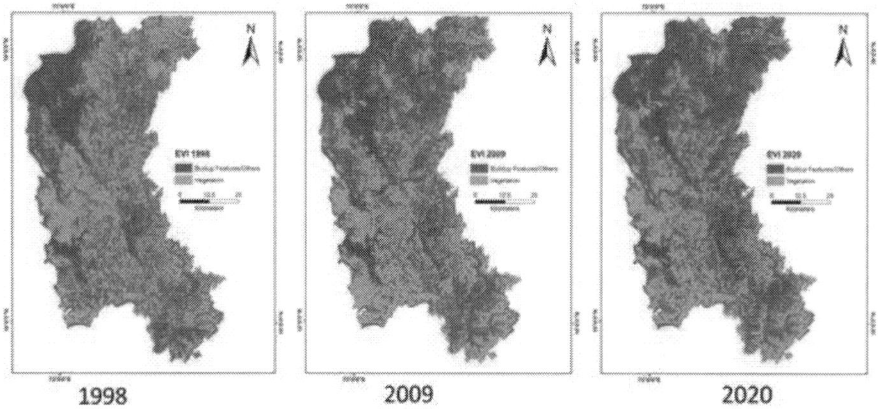

Figure 3.13: Enhanced Vegetative Index (EVI)

3.8.4.1 TM Image Interpretation (1998) for NDVI

To find the vegetation index, change detection analysis, various maps, Figures, and Tables were generated. It has been observed that approximately 54.9 % of the land, i.e., 3932.0514 hectares, is covered with vegetation. Moreover, approximately 45.1 % of the land has shown urbanization in 1998. Comparison of Landsat satellite images by 2009 TM with 1998 TM data set is shown in Figure 3.14, where the built-up area is shown in orange color and vegetation is shown in green color respectively.

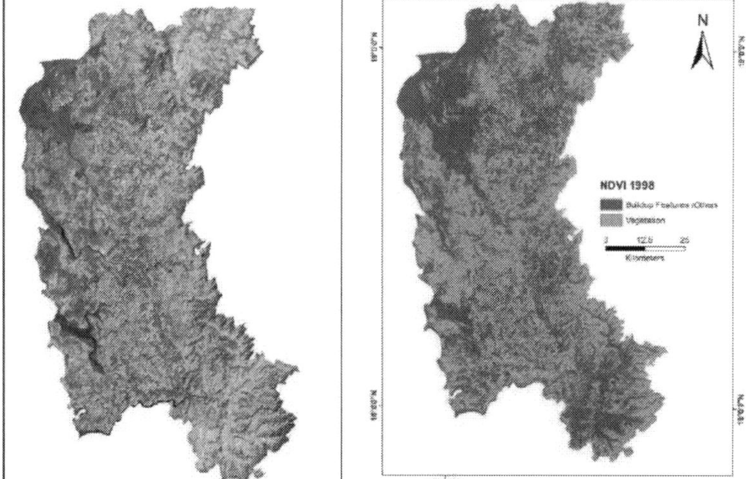

Figure 3.14: TM Image interpretation (1998) showing NDVI Map of 1998

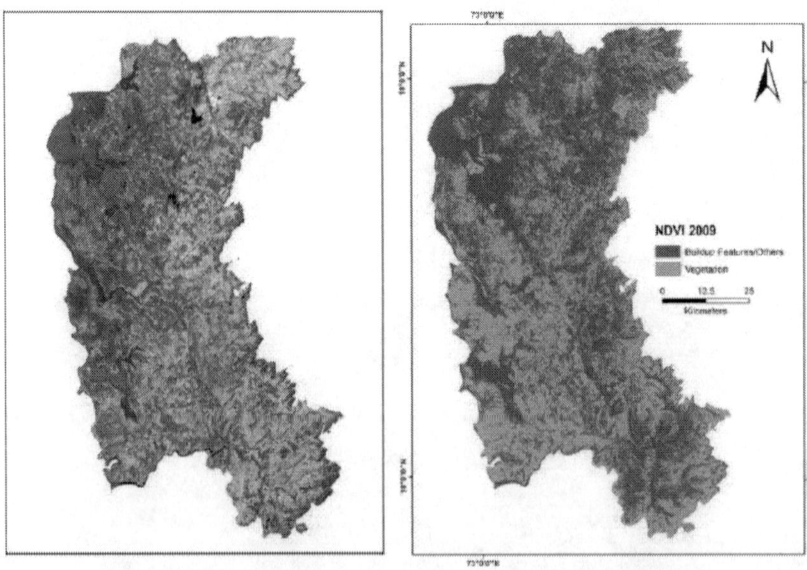

Figure 3.15: TM Image interpretation (2009) showing the NDVI Map of 2009

3.8.4.2 TM Image Interpretation (2009) for NDVI

Over the past 11 years (since 1998), as shown in Figure 3.15, rapid development happened, which decreased the rate of green vegetation in the district. Over 3512.2851 hectare, approximately 49.1 % of the land was covered under the built-up area. The rate increased by 4.0 % as compared with the 1998 data. In contrast, the vegetation was decreased because of industrialization and rapid growth in urban areas.

3.8.4.3 Operational Land Imager (OLI) Image interpretation (2009) for NDVI

Compared to the 1998 data, over 20 years, as shown in Figure 3.16, vegetation was decreased by 9.8 %, which is 698.78 hectares of area. Since 2009, analysis shows that the vegetation was decreased by 5.7 %, i.e., 411.3648 hectares of area, over the last ten years.

As shown in Figure 3.16, the change detection was done from 1998 to 2020 with 11 years interval. A clear view of green vegetation and built-up features were generated, and the results are tabulated in Table no 3.9.

Table 3.9: Normalized Difference Vegetative Index (NDVI) Dynamics

NDVI Statics	Area in Sq.Km	
YEAR	**Green Vegetation**	**Builtup/ Other features**
1998	3932.0514	3224.8494
2009	3644.6166	3512.2851
2020	3233.2518	3923.6013

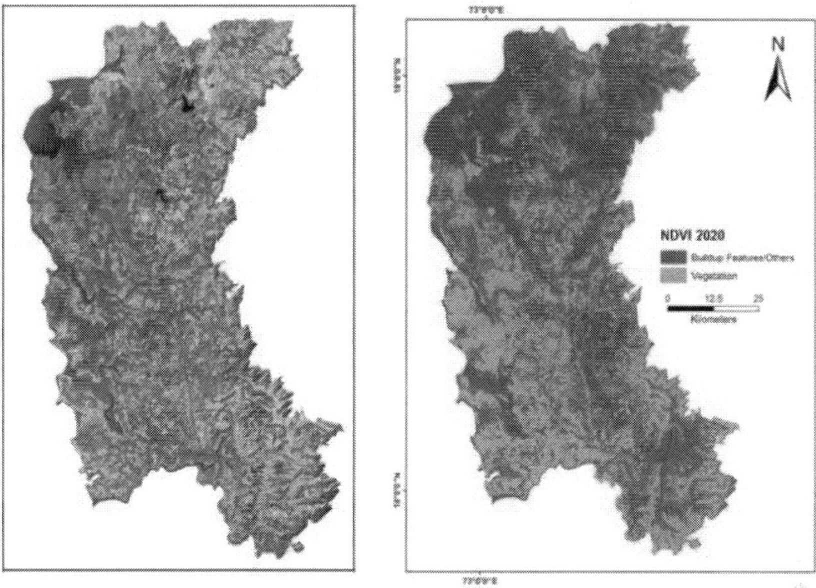

Figure 3.16: OLI Image interpretation (2009) showing the NDVI Map of 2020

3.8.4.4 TM Image interpretation (1998) for EVI

To find the vegetation index, change detection analysis, various maps, Figures, and Tables were generated. It has been observed that approximately 54.9 % of the land, i.e., 3919.9 a hectare is covered with vegetation, and approximately 45.3 % of the land has shown urbanization in 1998. Comparison of Landsat satellite images by 2009 TM with 1998 TM data set is shown in Figure.3.17, where the built-up area is shown in orange color and vegetation is shown in green color respectively.

3.8.4.5 TM Image Interpretation (2009) for EVI

Over the past 11 years (since 1998), as shown in Figure 3.15, rapid development happened, which decreased the rate of green vegetation in the district. Over 3444.2 hectare approximately 48.1 % of the land was covered under the built-up area. The rate increased by 2.8 % as compared with the 1998 data. In comparison, the vegetation was decreased because of industrialization and rapid growth in an urban area. Refer Figure.3.18

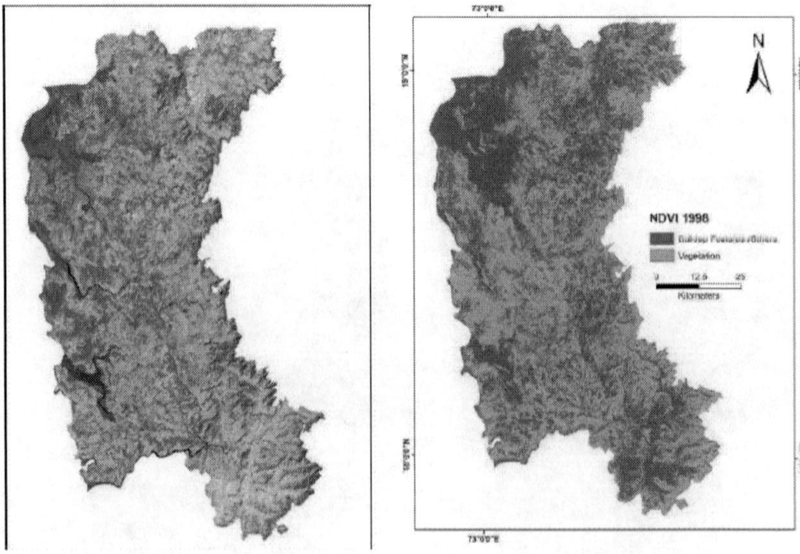

Figure 3.17: TM Image interpretation (1998) showing EVI Map of 1998

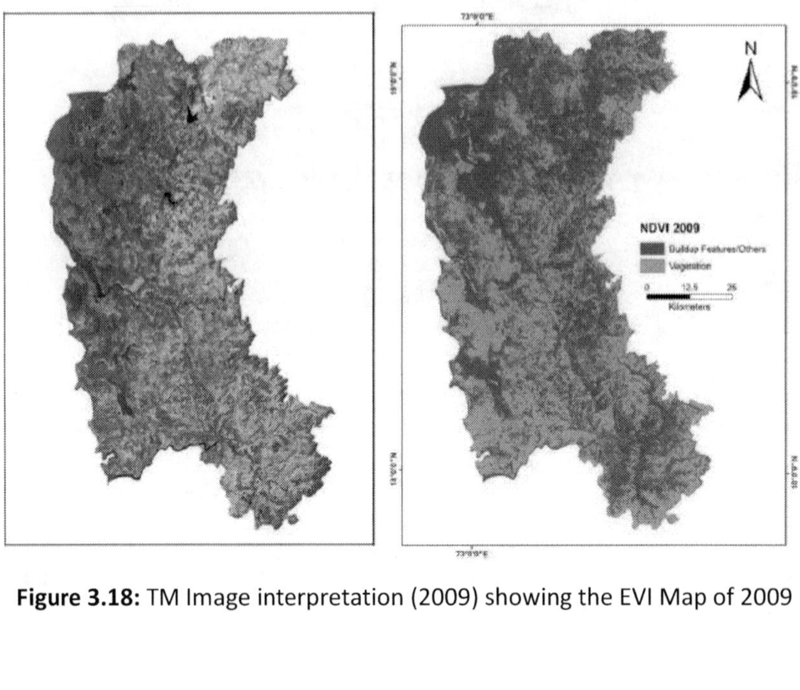

Figure 3.18: TM Image interpretation (2009) showing the EVI Map of 2009

3.8.4.6 Operational Land Imager (OLI) Image interpretation (2009) for EVI

Compared to the 1998 data, over 22 years, as shown in Figure 3.19, vegetation was decreased by 6.8 %, which is 698.78 hectares of area. Since 2009, analysis shows that the vegetation was decreased by 4.0 %, i.e., 282.79 hectares of area, over the last 11 years.

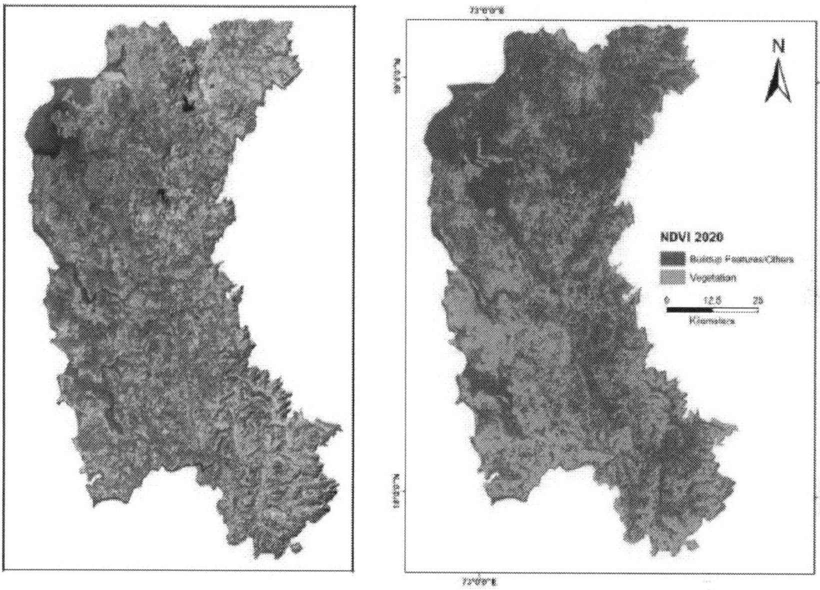

Figure 3.19: OLI Image interpretation (2009) showing EVI Map of 2020

As shown in Figure 3.19, the change detection was done from 1998 to 2020 with 11 years interval. A clear view of green vegetation and built-up features were generated, and the results are tabulated in Table no 3.10

Table 3.10: Enhanced Vegetative Index (EVI) Dynamics

EVI Statics	**Area in Sq.km**	
YEAR	**Green Vegetation**	**Builtup/ Other features**
1998	3913.9146	3242.9871
2009	3712.7322	3444.1686
2020	3429.9333	3726.9684

3.9 DIGITAL ELEVATION MODEL (DEM)

3.9.1 General

Digital Elevation Model DEM is defined as a 3-dimensional computer graphical technique of representation the earth's surface. Every point on the earth's crust has its own identity regarding geographical parameters such as latitude, longitude and altitude. These parameters vary and differ, and hence the whole globe, when seen from a bird's eye view, is undulated with terrains and valleys, which creates its profile geometrically.

Figure 3.20: Satellite Data acquired for DEM Generation
(**Source:** www.usgs.gov .in)

As shown in Figure 3.20, Remote Sensing (ResourseSat and CartoSat Satellite image (Bhuvan, ISRO's portal) and GIS techniques (3D DEM) were applied to simulate DEM. Hence this research required a 3-dimensional map to visualize the ground slope so that the WTN and WDN can be laid considering the gravity hence optimizing the usage of pumps which plays a vital role in planning and executing the overall budget for WTN infrastructural projects.

3.9.2 Georeferencing

As discussed above in the objective, the concept of georeferencing is to coincide the satellite data (spatial data) with the toposheet (non-spatial data) of the region under research.

3.9.3 Georeferencing and False Colour Composite

To summarize, this research work signifies that the process of georeferencing is thoroughly described in a layman way, from the procedure of how to freely download Toposheet from Survey of India to downloading LISS-III satellite data from ISRO's Bhuvan portal. The step- by-step procedure of performing the georeferencing technique in QGIS, open-source software using various tools, is implemented and explains all necessary technical and scientific details. This study finally concludes an accurate georeffered satellite imagery pinned over the toposheet at high accuracy by performing plate spline transformation technique using the nearest neighbour technique with resampling method as SRS target as EPSG 3857 WGS 84. Toposheet and False Color Composite of the study area is shown in Figure 3.21.

The results emphasize that all the four courses of the toposheet and the LISS III imagery get superimposed with 18°.58'.44.61" to 19°.0'.57.16" latitude and 73°.02'.54" to 73°.05'.39.61" longitude respectively of the study area. It can be concluded that this rereferred out data can be further used for finding various GIS applications such and vegetative cover change dynamics analysis, network analysis, shapefiles, which increases the percentage of accuracy in the GIS.

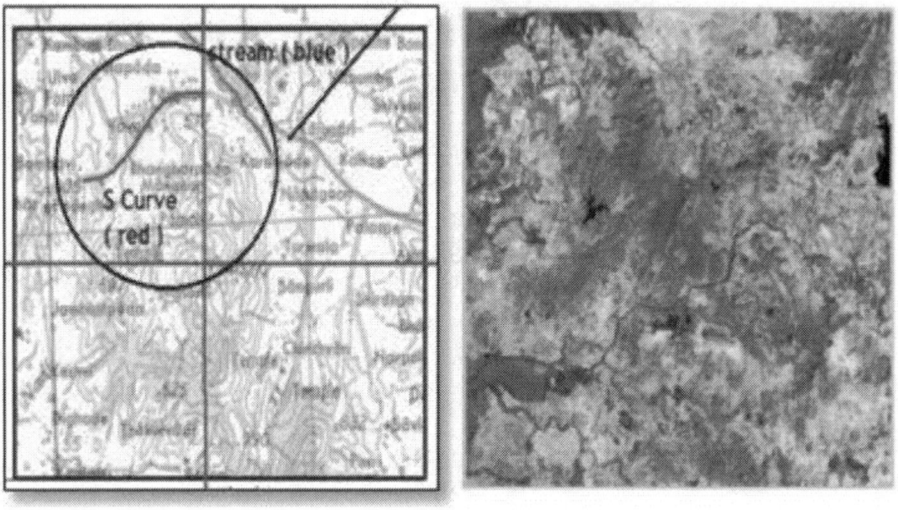

Figure 3.21: Toposheet & FCC of the study area
(**Source:** www.soi.in)

The georeferred maps help in various GIS applications, as shown in Figure 3.21, such as generating DEM and bringing the accuracy level coinciding with WTN and WDN in the correct route. It has been seen that through FCC, more clear visualization is achieved hence helping this research to assume the suitable routes take correct decisions for urban planning.

3.9.4 Digital Elevation Model

A simulated DEM (Figure 3.22) is generated, and the results show that a well-planned WTN/WDN can not only optimize the network through DEM but also save water to a great extend (as the ground slope is known after simulation)if hydraulic parameters such as velocity, head, discharge, are designed within its permissible limits considering the water demand. The lowest and highest elevation which we got are 3.8 m and 43.8 m from the mean sea level in DEM.

Figure 3.22: DEM simulation of the study area

As seen in Figure 3.23, LISS III satellite imagery is downloaded free from IROS's Bhuvan portal, and the toposheet is downloaded free from Survey of India by providing the latitude and longitude coordinates as stated in the previous paragraph. All the four bands are combined by a process termed Image staking. Refer to Figure 3.24.

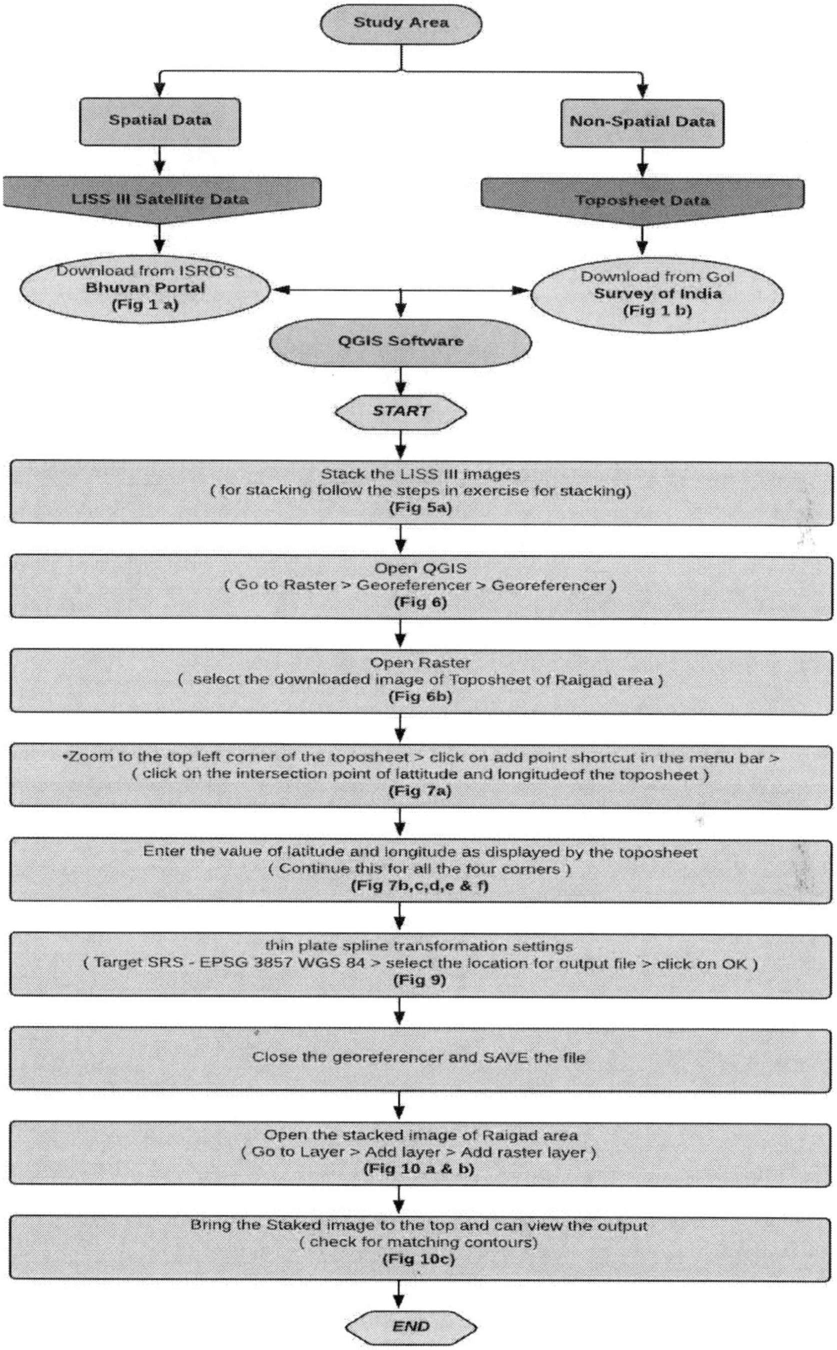

Figure 3.23: Methodology for Georeferencing

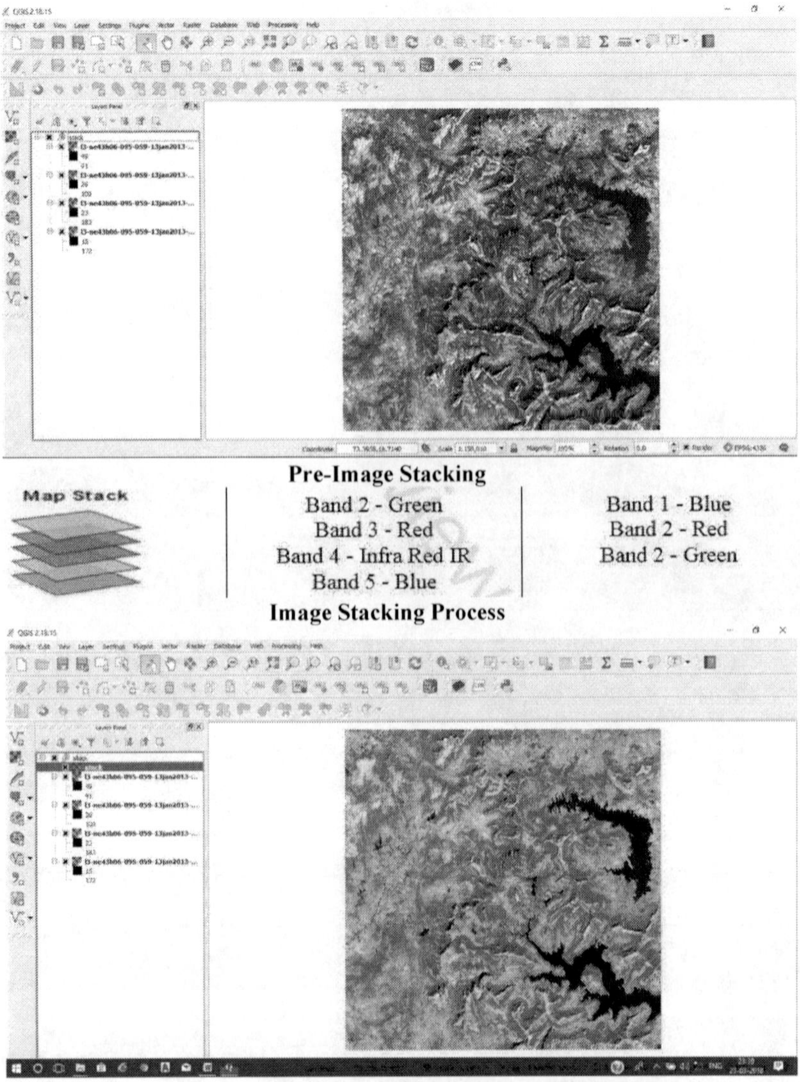

Post Image Stacking

Figure 3.24: Pre and Post Image stacking LISS-III Satellite Imagery outputs

As mentioned in 1.3.1, the toposheet of India was downloaded from Survey of India www.surveyofindia.gov.in, and in that state of Maharashtra falls under NE-43 sheet number, which is later subdivided into 16 sub numbers as shown in Figure 5 b. In that Mumbai and Navi Mumbai lies in mini block 1 and 5 respectively. As in this research, the study area is the Raigad region which also lies in mini-block no 5, the toposheet number selected for this research is NE-43-5. A zoomed image of the study area is also inserted in Figure 3.25, which shows a red color S curve, national highway number 348, also known as JNPT road, which lies in the study area.

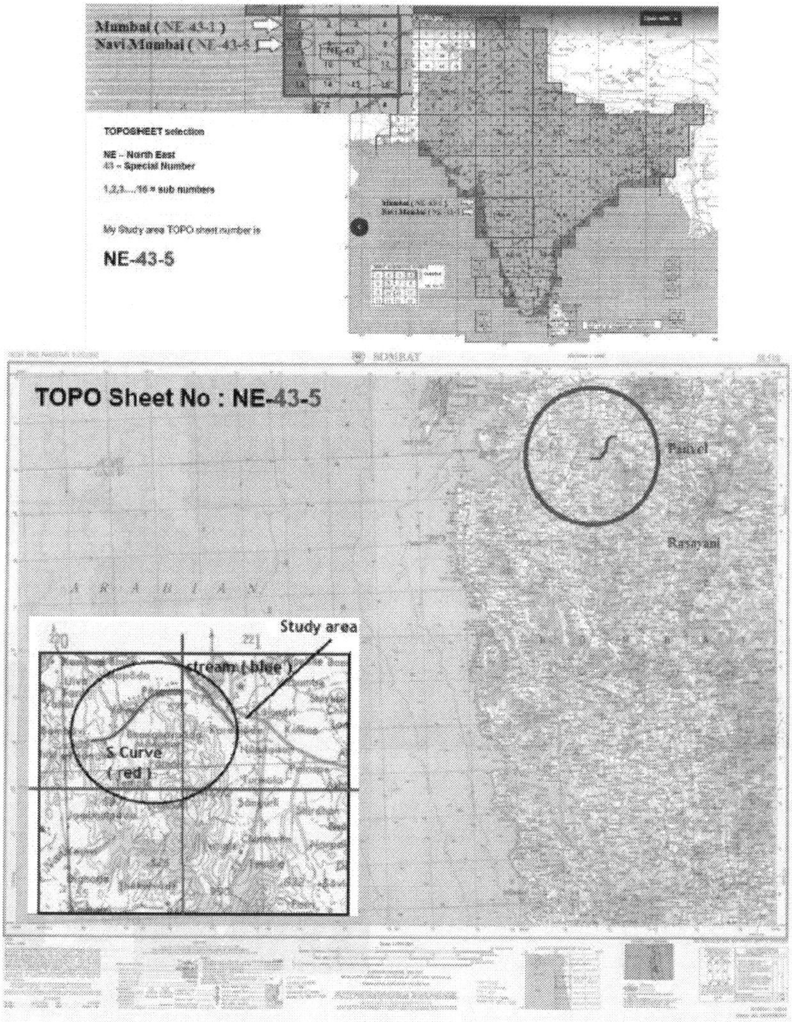

Figure 3.25: Toposheet of the study area and its identification numbering pattern (Red color S curve, which is national highway number 348 - JNPT road

The raster image (are digital photographs representing spectral data) is then opened in QGIS software in which a tooled name Georeferencer is selected. Later the toposheet is downloaded from the Survey of India portal. Refer to Figure 3.26.

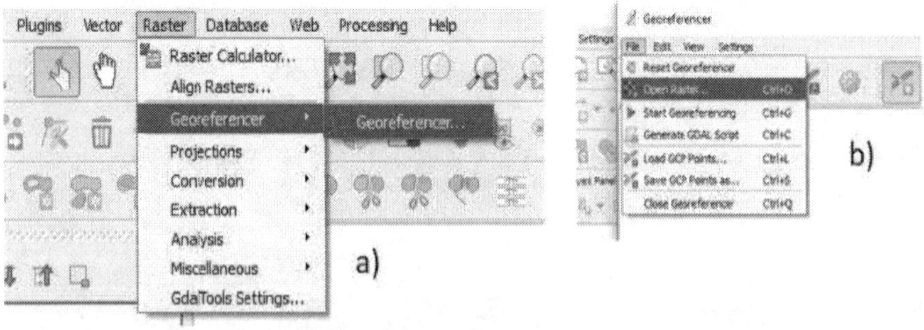

Figure 3.26: QGIS application showing Georeferencer and raster tool for Georeferencing. a) Georeferencer tool b) Raster tool in QGIS

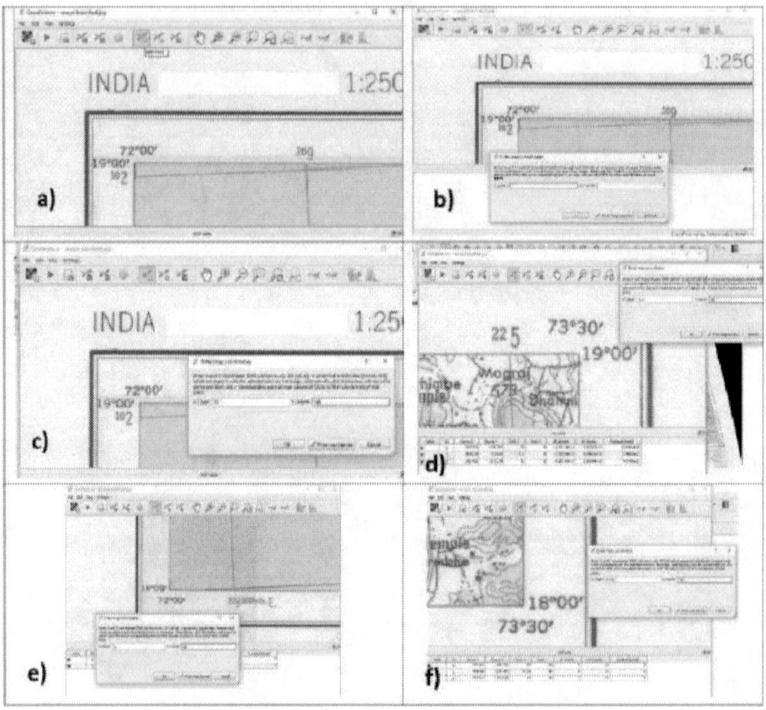

Figure 3.27: Assigning latitude & longitude of study are overall the four corners of the toposheet using intersection tool in QGIS

After selecting the toposheet, in the left corner is zoomed in, the intersection of the latitude (190 00') and longitude (1720 00') can be seen. (Refer to Figure 7a). Enter these latitude and longitude values in the Table, which pops up once clicked at the intersection as shown in Figure 7b and c. Continue this same process of entering the latitude and longitude values in all the remaining three corners of the toposheet, respectively, as shown in Figure 3.27 d, e, and f respectively. The features that can be seen over the toposheet or the raw satellite imagery for which we have demarcated known ground coordinates are termed Ground Control Points (GCPs) in GIS terminology. These GCPs connects a relation between the ground and (toposheet or raw satellite imagery) by correlating the pixel by P and line by L coordinates on the image with the ground coordinates x, y, and z, respectively, as shown in Table 3.11

Table 3.11: Ground Control Points (GCPs) of the study area

GCP table

Visible	ID	Source X	Source Y	Dest. X	Dest. Y	dX (pixels)	dY (pixels)	Residual (pixels)
✖	0	393.782	-258.531	72	19	0	0	0
✖	1	4595.68	-256.437	73.25	19	0	0	0
✖	2	382.824	-3212.93	72	18	0	0	0

The technique by which data are interpolated and smoothened to geometric design was introduced by (Duchon 1976), generally termed Thin-Plate Splines (TPS).

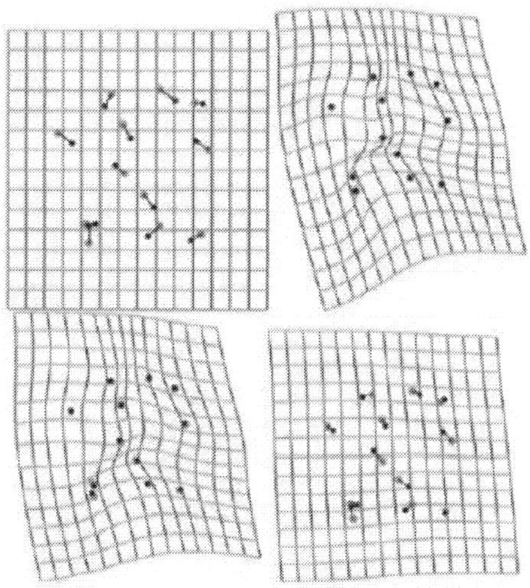

Figure 3.28: Thin-Plate Splines (TPS) representation techniques. (Source: DOI: 10.1109/IEMBS.1996.652767)

These TPS are generally used for representing the transformation of shapes for matching and detection applications, as shown in Figure 3.28. In QGIS Thin Plates Splines TPS is provided in the transformation settings wherein the nearest neighbor technique is considered for resampling method by setting SRS target as EPSG 3857 WGS 84 as shown in Figure 3.29.

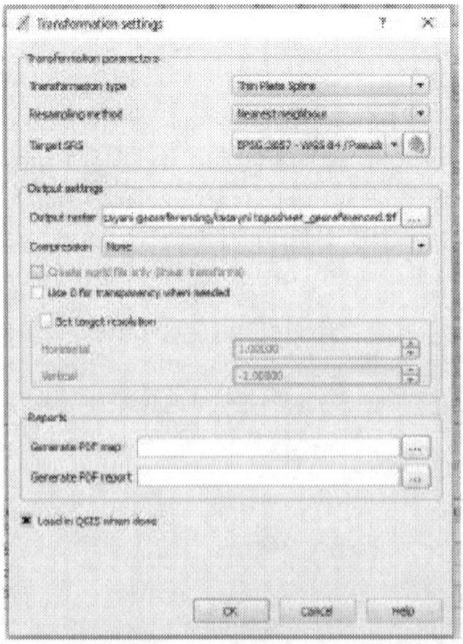

Figure 3.29: QGIS toolbox showing Thin-Plate Splines (TPS).

After closing the Georeferencer, the stacked satellite image of the Raigad region is added in the Raster layer, as shown in Figure 3.30 , which later can be adjusted and brought to the top of the toposheet for proper visualization, as shown in Figure 3.30.

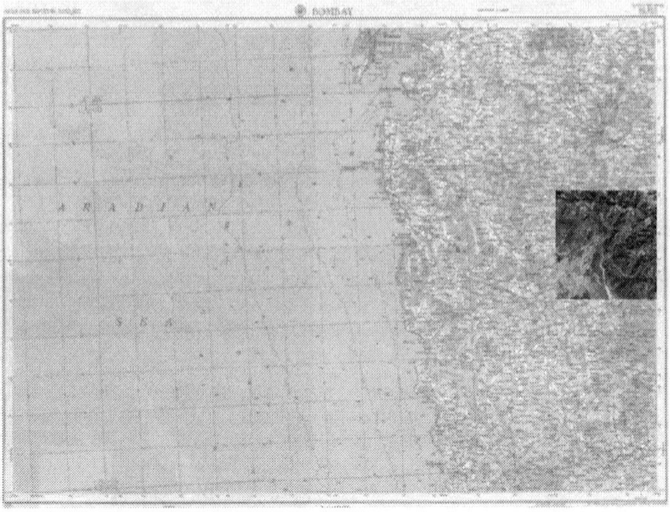

Figure 3.30: Final result showing the georeffered Raigad toposheet pined over LISS-III satellite imagery.

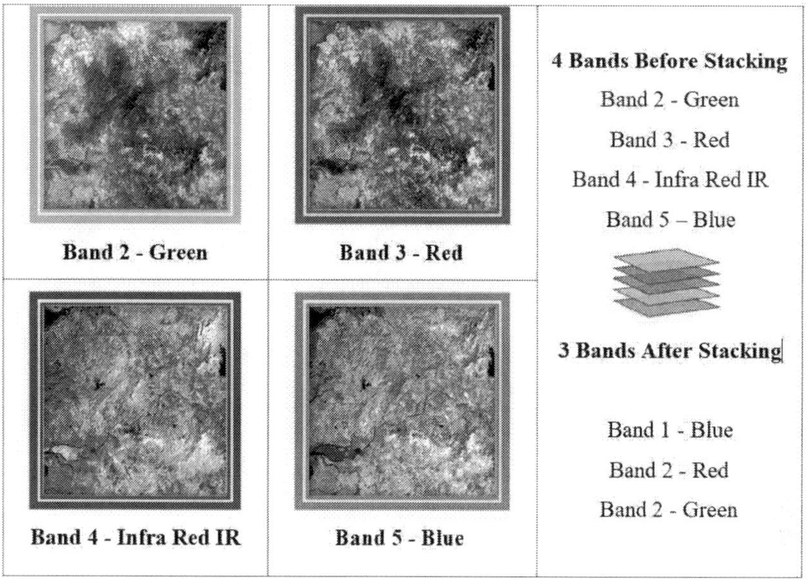

Figure 3.31: False Color Composite of the study area prepared in QGIS from ResourseSat I LISS III (Band 2,3,4 and 5) and CARTOSAT I Satellite images by Layer Stacking

3.10 FALSE COLOUR COMPOSITES (FCC)

3.10.1 General

Color Composite images are defined as satellite images exhibited in 3 primary basic Colours, red, green, and blue (RGB), associating spectral bands in remote sensing and are distinguished as True Colour Composites and False Colour Composites (FCC).

FCC helps in understating the satellite image through visualization as in wavelengths in bands like near-infrared and beyond, which the human eye cannot see. These four bands R, G, B, and IR are superimposed (compressed) over each other, which is termed layer stacking, as shown in Figure 3.31 later helps in RS and GIS applications.

3.10.2 Procedure for Stacking of Satellite Images in QGIS

For layer stacking, the Resourcesat I: LISS III images which are in raster format, are arranged according to band numbers 2,3,4 and 5 grouped to form an output file which is in tiff format (Tagged Image Format File.) which would be later required to be overlapped Cartosat I DEM file to get the elevation profile of the study area. Layer stacking procedure is explained in a hierarchy format in Figure 3.32

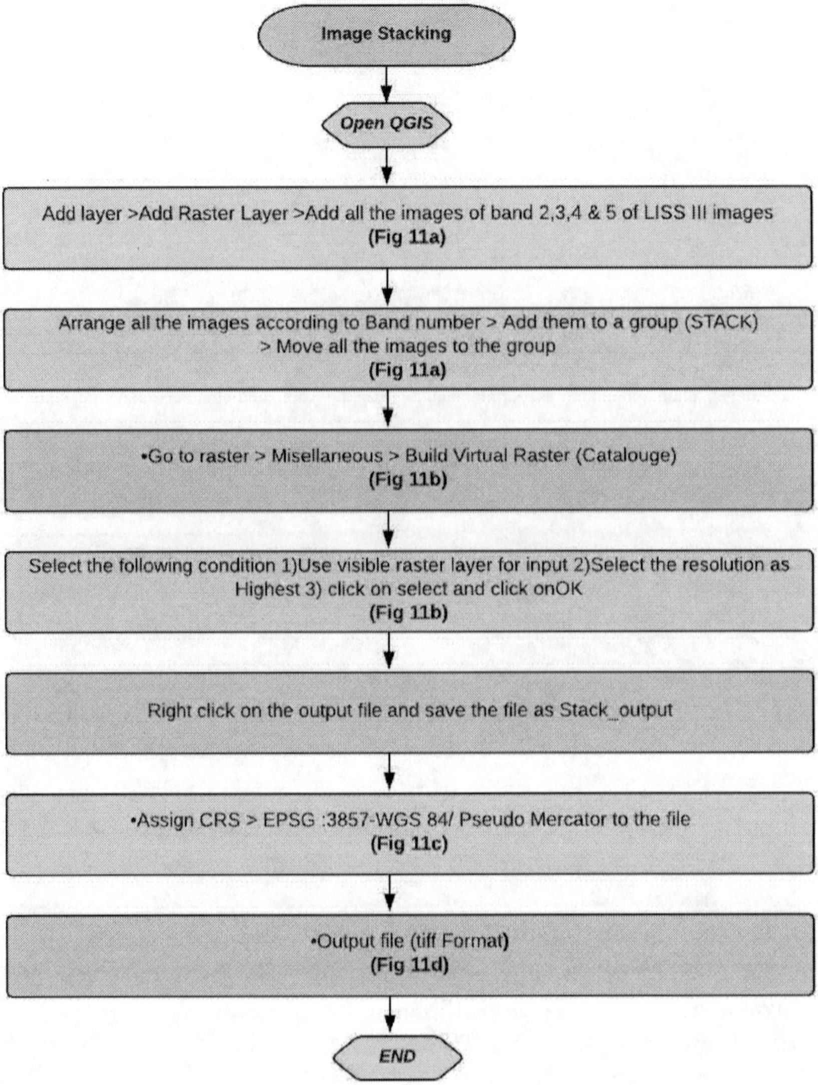

Figure 3.32: Satellite Image stacking procedure

All the LISS-III images of bands 2,3,4, and 5 are added to the raster layer and arranged according to band numbers, and converted into groups, i.e., Stack using Build Virtual Raster catalog visible raster layer selected with the highest resolution and programmed ahead for the stacked output file. As mentioned above, EPSG-3857-WGS 84/ Pseudo Mercator is assigned as the Central reference System (CRS), and the output is obtained in TIFF format. The procedure can be seen in Figure 3.33.

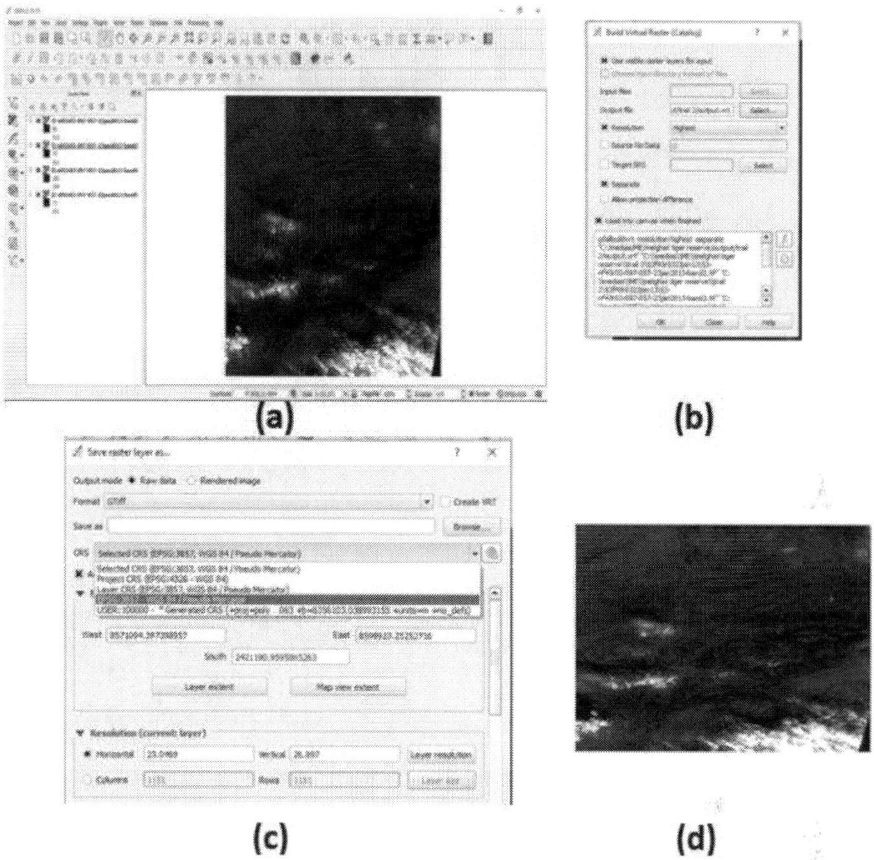

Figure 3.33(a): Band 2,3,4 and 5 of LISS III satellite image added together (b) highest resolution (c) Assigning CRS EPSG:3857-WGS 84/ Pseudo Mercator to the file (d) Output File

3.10.3 False Color Composite

When bands or maps are created by combining three raster images, a True Color Composite is formed. These bands are displayed in three primary colors red, green, and blue. These color composites are created for illustration or easy visualization purposes, and they can also be overlapped over a three-dimensional model. A composite that is prepared by an arbitrary technique by assigning any multispectral band where the actual color of the target does not resemble the displayed color on the image is termed as False Color Composite (FCC). Refer to Figure 3.34.

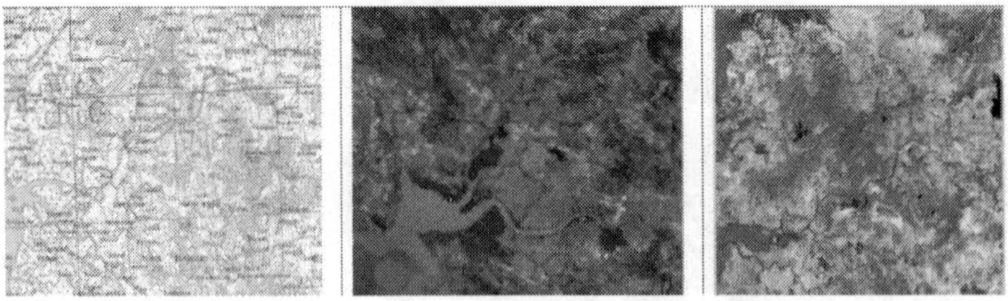

Figure 3.34: Toposheet, True Color Composite, and False Color Composite of Regard

The wavelengths in the near-infra-red region of the electromagnetic spectrum are not visible to the human eye, and hence FCC is a representation of MSS images using image components of RGB instead of visible RGB.

In LISS-III, the band combinations are Band 2 varies from 0.52 - 0.59 microns, Band 3 varies from 0.62 - 0.68 microns, Band 4 varies from 0.77 - 0.86 microns, and Band 5 varies from 1.55 - 1.70 microns, respectively.

3.10.4 Creating False Color Composite (FCC)

After unzipping the LISS III Multispectral satellite MSS image, which contains 4 TIFF files, i.e., band 2,3 4, and 5 (which has 24 m spatial resolution ortho-rectified from IRS Resourcesat satellite series)

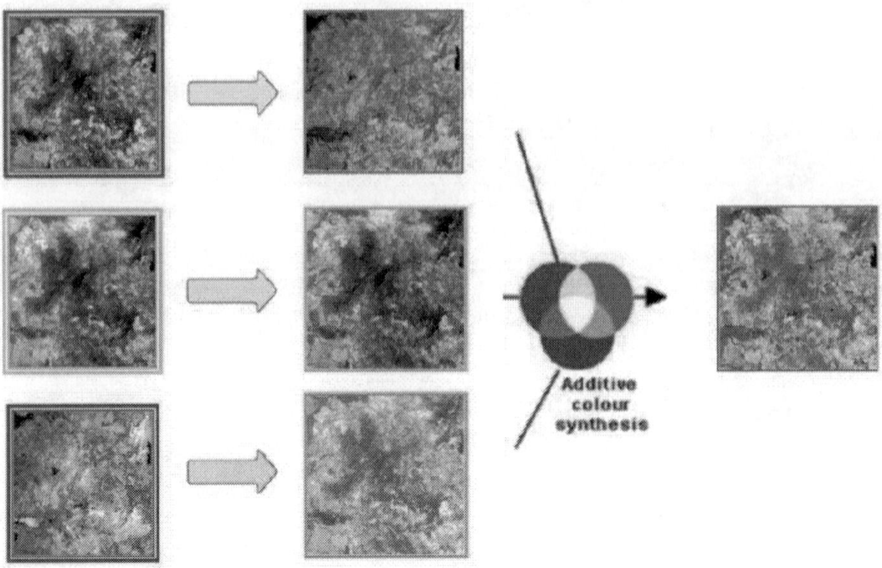

Figure 3.35: Layer staking technique for generating FCC

After downloading all four bands of the image and stacking, the output file is converted into False Color Composite FCC, which helps to understand the topography of the study area. Figure 3.35 shows the layer staking, which helps to understand the topology and other spatial parameters.

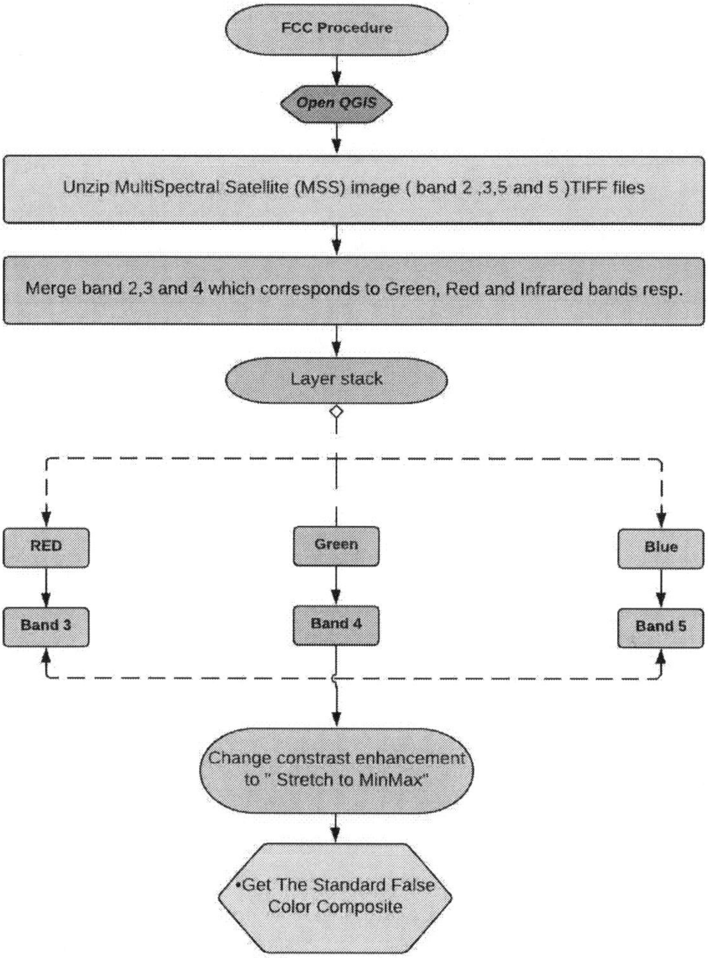

Figure 3.36: Procedure for creating False-color composite FCC Image

These layers are later geo-referenced to get an accurate understanding of latitude and longitude. After georeferencing is done, the digital elevation model (DEM) is generated in QGIS, which helps to understand the ground slope, which helps to lay the water distribution networks in the direction of gravity, reducing the introduction of pumps. Refer to Figure 3.36.

<table>
<tr><td colspan="3">**Resoursesat I: LISS III downloaded image (Band 2,3,4 and 5)**</td></tr>
<tr><td colspan="2"></td><td></td></tr>
<tr><td colspan="2">**Band 2 – Green**</td><td>**Band 3 - Red**</td></tr>
<tr><td colspan="2"></td><td></td></tr>
<tr><td colspan="2">**Band 4 - Infra-Red IR**</td><td>**Band 5 - Blue**</td></tr>
<tr><td>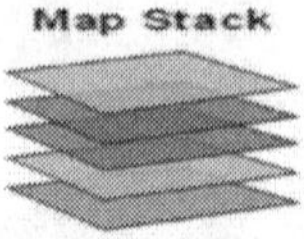</td><td>Band 2 - Green
Band 3 - Red
Band 4 - Infra-Red IR
Band 5 - Blue</td><td>Band 1 - Blue
Band 2 - Red
Band 2 - Green</td></tr>
<tr><td>Layer Stack (LS)</td><td>Before LS</td><td>After LS for preparing FCC</td></tr>
<tr><td colspan="3"></td></tr>
<tr><td colspan="3">**False Color Composite FCC of the Study area**</td></tr>
</table>

Figure 3.37: False Color Composite of the study area prepared in QGIS from Resourcesat I: LISS III downloaded image (Band 2,3,4 and 5) images by Layer Stacking

Refer to Figure 3.37. The display color assignment for any band of a multispectral image can be done in an entirely arbitrary manner. In this case, the color of a target in the displayed image does not resemble its actual color. The resulting product is known as a false-color composite image. There are many possible schemes for producing false-color composite images. However, some schemes may be more suiTable for detecting particular objects in the image. For example, a widespread false-color composite scheme for displaying a SPOT multispectral image is shown below:

R = XS3 (NIR band), G = XS2 (red band), B = XS1 (green band). This false-color composite scheme allows vegetation to be detected readily in the image. In this type of false-color composite image, vegetation appears in different shades of red depending on the types and conditions of the vegetation since it has a high reflectance in the NIR band (as shown in the graph of spectral reflectance signature). Clearwater appears dark-bluish (higher green band reflectance), while turbid water appears cyan (higher red reflectance due to sediments) than clear water. Bare soils, roads, and buildings may appear in various shades of blue, yellow, or grey, depending on their composition (book: Photogrammetry, GIS & Remote Sensing- S S Manugula, Veeranna Bommakanti)

3.10.5 Details about Digital Elevation Model (DEM)

Digital Elevation Model DEM is defined as a 3-dimensional computer graphical technique of representation the earth's surface. Every point on the earth's crust has its own identity regarding geographical parameters such as latitude, longitude, and altitude. These parameters vary and differ, and hence the whole globe, when seen from a bird's eye view, is undulated with terrains, valleys, which creates its profile geometrically.

Remote Sensing (Resourcesat and CartoSat) Satellite image (Bhuvan, ISRO's portal) and GIS techniques (3D DEM) are applied and utilized to simulate DEM. Hence, this research required a 3-dimensional map to visualize the ground slope so that the WTN and WDN can be laid considering the gravity, hence optimizing pumps' usage, which plays a vital role in the overall budget for WTN infrastructural projects.

The traditional method of creating DEM is to join the elevation along with two mutual directions and then superimpose to simulate a 3 D model, as shown in Figure 3.38.

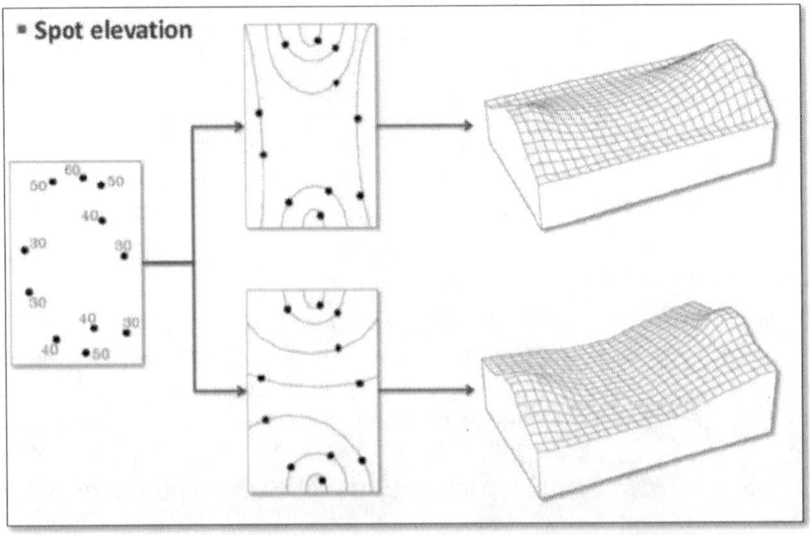

Figure 3.38: Traditional Method of DEM Generation
(**Source:** www.gisresources.com)

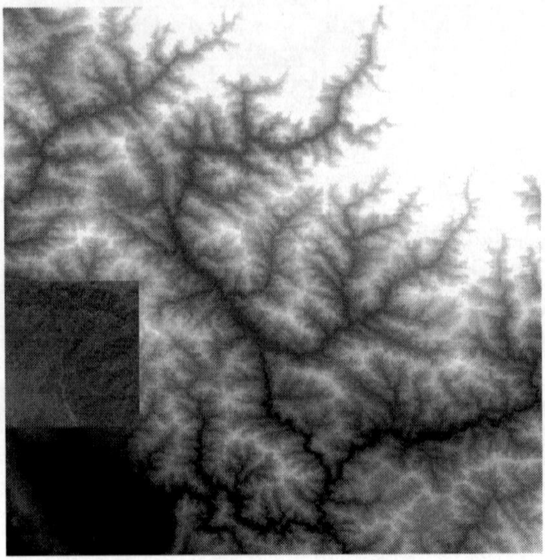

Figure 3.39: Cartosat-I DEM Satellite data of the study area

Figure 3.39 shows Cartosat-I DEM Satellite data of the study area. Digital Elevation Model (DEM) is also sometimes referred to as Digital Terrain Model (DTM) and Digital Surface Model (DSM). These DSM and DEM are two-dimensional maps where the altitude is of every point on the ground is information that is coded, whereas DTM is a numerical representation of the surface of the earth in 3 dimensions where every point of the terrain has its coordinates in x, y, and z-direction in digital format. Earth's surface is showing having buildings, trees, towers., if shown with their surface boundaries, is termed as DEM, and represented without boundaries, it is termed as DSM. Figure 3.40 shows the difference between DEM and DSM.

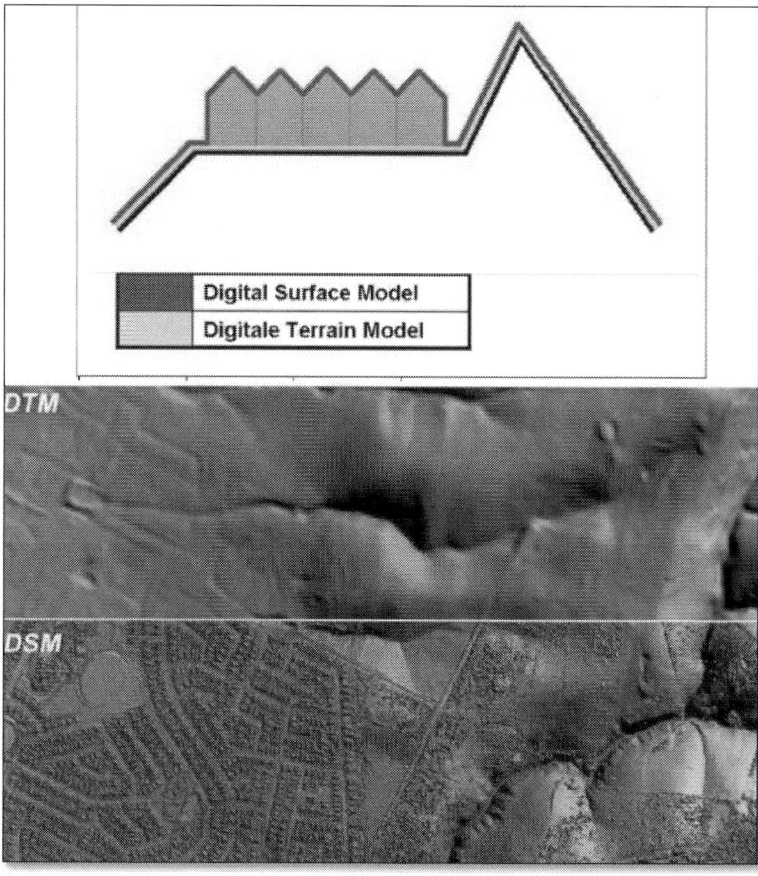

Figure 3.40: DEM and DTM
(**Source:** www.gisresources.com)

3.10.6 Creating 3D model (DTM) from Satellite Image

The methodology of generating 3D DEM in QGIS is shown in a flowchart, as shown in Figure 3.41. shows the explanation of this procedure.

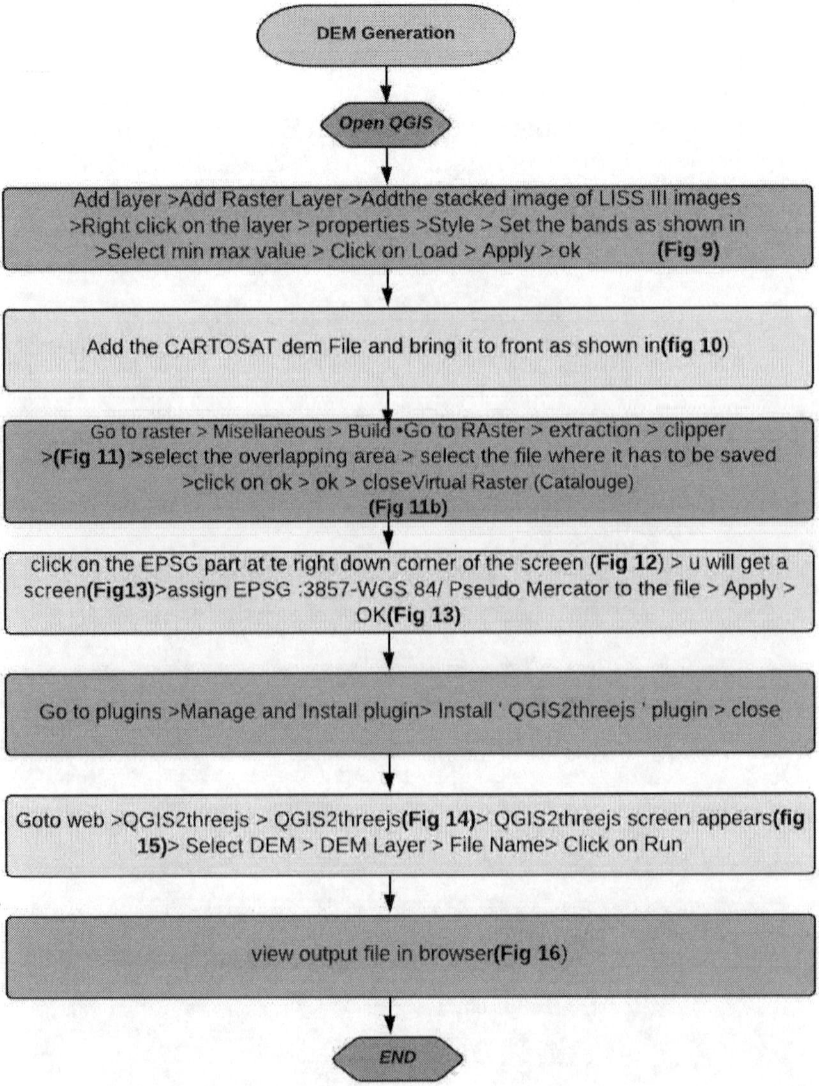

Figure 3.41: Methodology for DEM generation

For generating 3d Model by DEM in QGIS, initially, that stacked images of LISSS_III are added together as a raster layer wherein the band combination properties are set as shown in Figure 3.42a minimum and the maximum value is loaded to it. Cartosat-I DEM file is superimposed as shown in Figure 3.42b over LISS-III stacked imagery using a clipper tool as shown in Figure 3.42c. EPSG:3857 84/Pseudo Mercator (refer Figured and 3.42d) is considered the datum and QGIS2threejs plugin installed (Figure 3.42e). When this file is open on the web with an Internet connection, a 3-dimensional model is simulated in the output file, as shown in Figure 3.42.

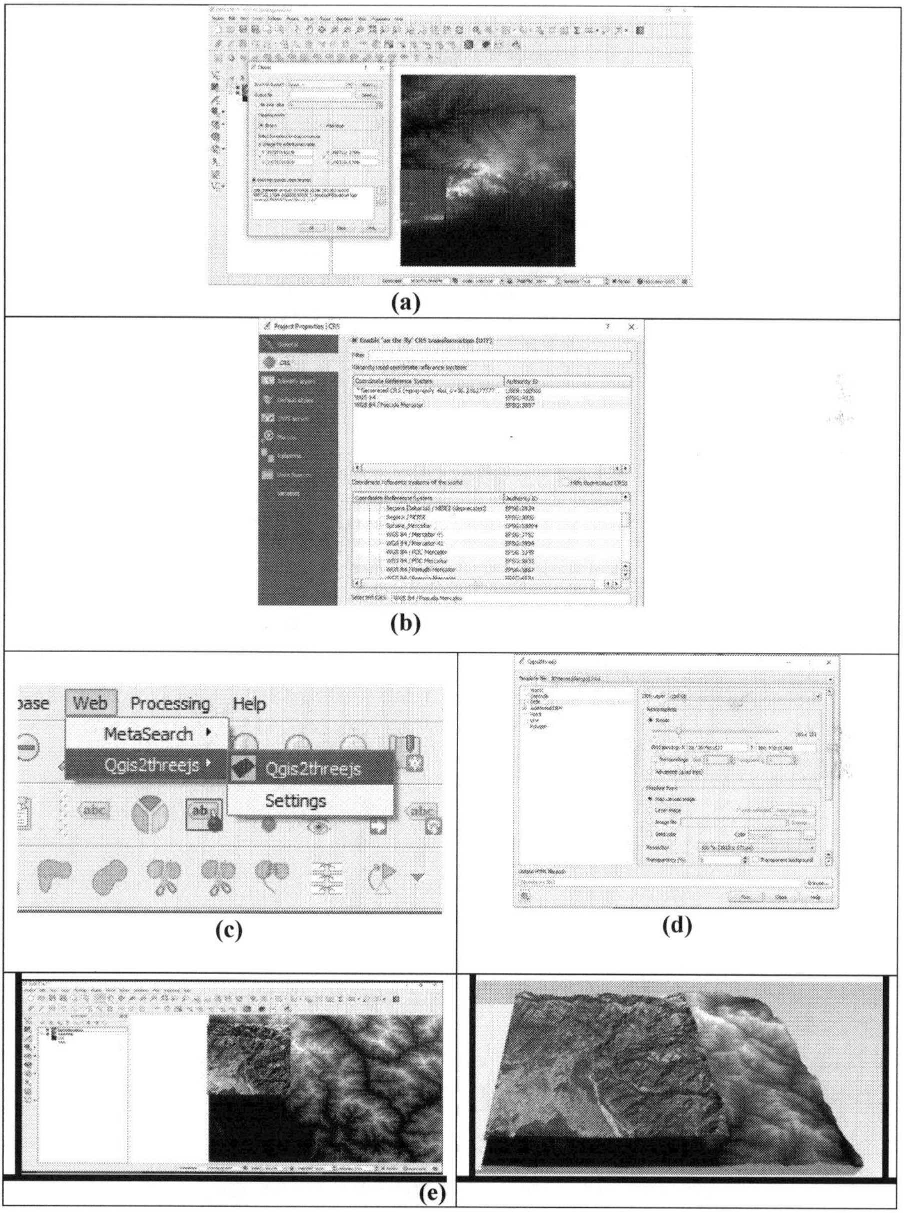

Figure 3.42: DEM generation

3.11 WATER TRANSMISSION NETWORK (WTN)

City and Industrial Development Cooperation Ltd. (CIDCO) of Maharashtra is India's Premiere town planning agency, established in 1970 to plan and develop new towns. CIDCO has planned a water supply system for its area, and around 150 MLD at water treatment plant WTP has been constructed at Jite. The methodology explained ahead is to verify the design of the WTN from the Hetwane dam to the 150 MLD Jite (refer to Figure 3.42a). Presently a PSC pipeline has been laid to transmit water from the Hetwane dam to the sump and pump house to Jite. Water from this sump would be pumped to the WTP at Jite. During the passage from the dam to the sump and pump house, 20 MLD water is supplied to the five enroute nodes of the villages. The requirement of water at WTP at Jite is 150 MLD. Hence the present system should pass 170 MLD water.

Figure 3.42a: Hetwane Dam Satellite View
(**Source:** http://mountshastaecology.org)

We consider all the factors and design criteria such as pressure, velocity, discharge, friction factor, frictional losses, losses at bends, pipe materials, costs, junctions, different valves. As discussed earlier in the abstract, we are giving eight alternative alignments and comparing them to give the most feasible transmission line, potentially meeting the public water demands.

The pipeline was required to lay after the node J-21 (Eirwadi). Analysis was required to assess the passage of water with an adequate quantity of 150 MLD to the sump and pump house. Thus, various scenarios of the existing, in-progress, and future pipelines have been study.

A most advanced technology of Geographic Information System using QGIS (open-source software) and water network simulation technique has been used to prepare a GIS-based hydraulic model using EPANET (open-source software). In these techniques, a hydraulic model simulating the gravity WTN has been prepared.

The scope of work involves verifying the hydraulic design of the existing and proposed alternative alignment of the Hetwane raw water gravity main for a water drawl of 150 MLD at WTP Jite. The basic scenario simulates water supply from the gravity pipelines to the sump and pump house and then to WTP has been prepared as a first step.

3.11.1 Parametric Investigations through Hydraulic Model

A Water Transmission line is defined as any pipeline conveying raw or treated water from a source or remote storage facility to a treatment plant and distribution storage tank. The present water transmission network is needed to be analyzed and redesigned to meet future water demand. In this project, we analyze a present water transmission network laid from the source, i.e., Hetwane Dam to Jite, where the water treatment plant is located, as shown in Figure 3.43.

We use modern software such as EPANET, QGIS, and remote sensing and GIS approaches to design an efficient WTN with accuracy for this project.

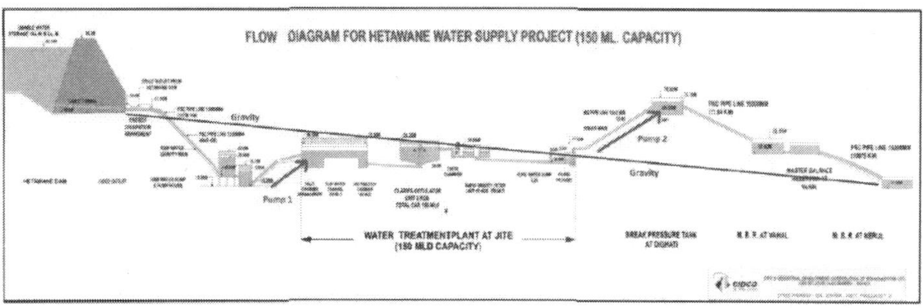

Figure 3.43: Water Transmission Network from the source to destination (**Source:** CIDCO)

Breakup shown in next page

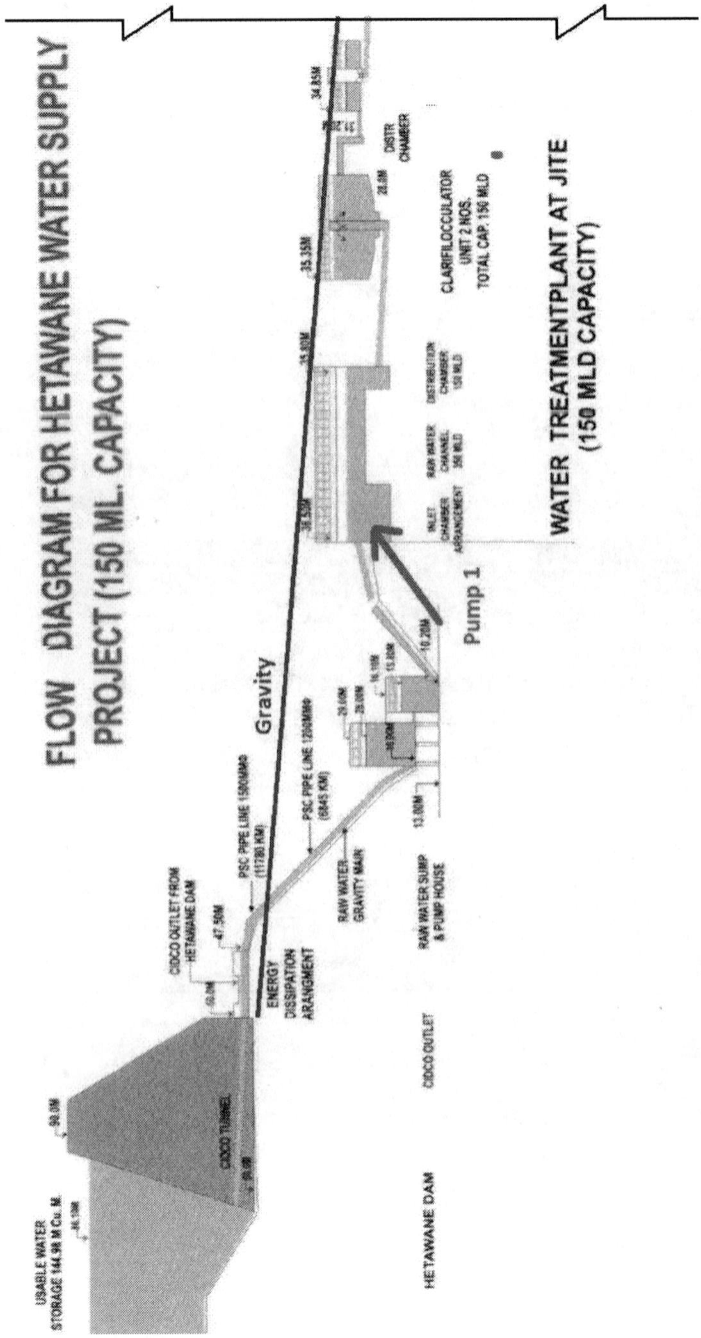
FLOW DIAGRAM FOR HETAWANE WATER SUPPLY PROJECT (150 ML. CAPACITY)
Gravity
Pump 1
WATER TREATMENTPLANT AT JITE (150 MLD CAPACITY)
CLARIFLOCCULATOR UNIT 2 NOS. TOTAL CAP. 150 MLD
ENERGY DISSIPATION ARANGMENT
RAW WATER GRAVITY MAIN
RAW WATER SUMP & PUMP HOUSE
CIDCO OUTLET
HETAWANE DAM
USABLE WATER STORAGE 144.98 M Cu. M

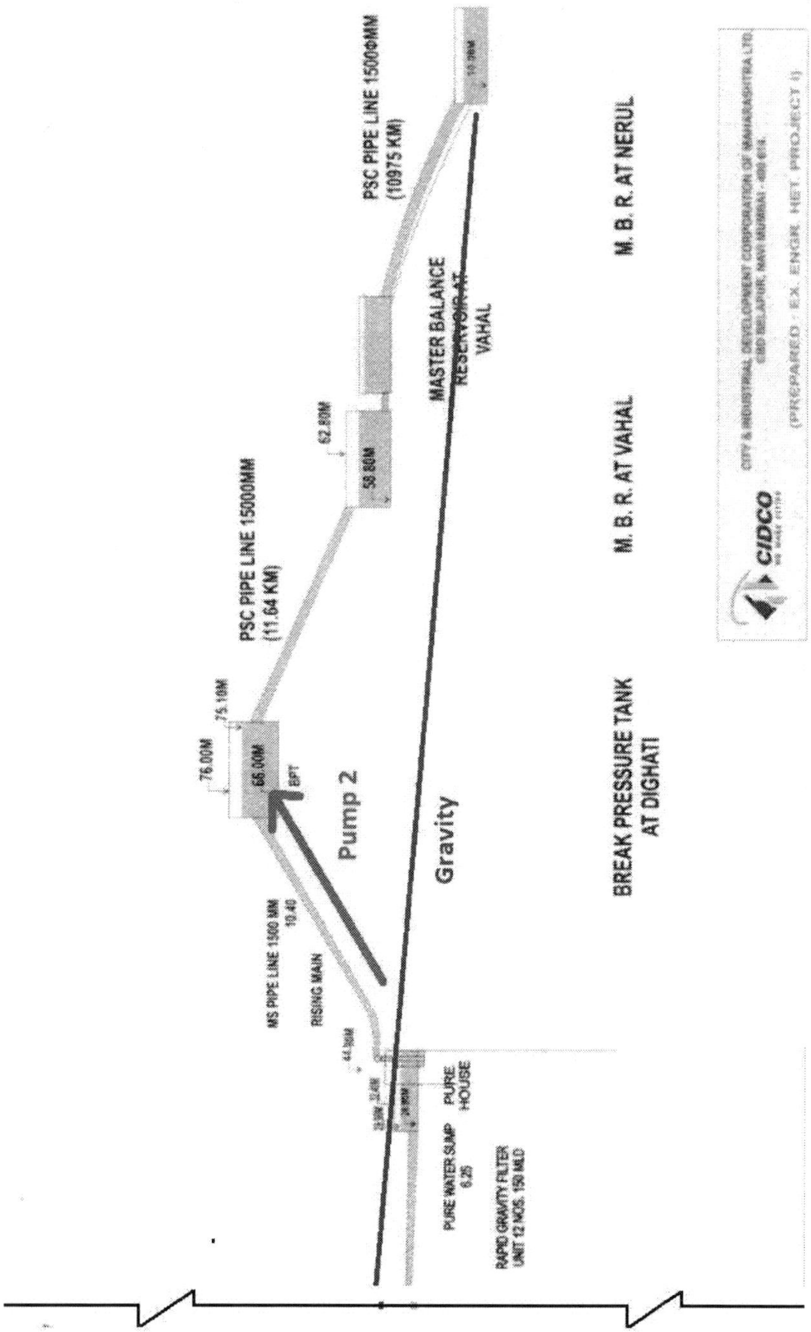
PSC PIPE LINE 1500ΦMM
(10975 KM)
M. B. R. AT NERUL
MASTER BALANCE RESERVOIR AT VAHAL
M. B. R. AT VAHAL
62.80M
58.80M
PSC PIPE LINE 15000MM
(11.64 KM)
75.10M
76.00M
66.00M
BPT
Pump 2
Gravity
BREAK PRESSURE TANK
AT DIGHATI
MS PIPE LINE 1500 MM
10.40
RISING MAIN
PURE HOUSE
PURE WATER SUMP
6.25
RAPID GRAVITY FILTER
UNIT 12 NOS. 150 MLD
CIDCO
(PREPARED : EX. ENGR. NET. PROJECT I)

A) Simulation model

Modeling the water supply system is critical for designing and operating water networks for a 27 x 7 continuous water supply. Hydraulic models give commanding knowledge of the water infrastructure and help to make informed decisions.

B) The Modeling process

The main objective of any water unity is to deliver safe and poTable water to its customers uninterruptedly. The first step in preparing the model of the water supply project is the creation of maps and records. System maps of the CIDCO area are made available by CIDCO. These maps helped a clear understanding of the transmission main. The maps illustrate a wide range of system characteristics such as pipeline alignment, the elevation of nodes, location of tanks, reservoirs.

Projected Population Vs. Water Supply Demand 2030

Table 3.12 explains the calculation of the projected population versus the water supply-demand in 2030. It can be seen that the overall water demand at 12 nodes is 700 MLD, and the water demand and consumption in the year 2014 are 210 MLD and 190 MLD, respectively. It can be seen in Table 3.10 that there is an incremental water demand at each of every 12 nodes from the year 2015 to 2030 based on urban sprawl by NDVI and EVI calculations. The total water supply-demand was 229 MLD, 388 MLD, 569 MLD, and 700 MLD in 2015, 2020, 2025, and 2030.

Deduction

30 % of water supply is considered to deduct losses and commercial consumption to derive population, which comes out to be 68.7 MLD, 116.4 MLD, 170.7 MLD, and 210 MLD in 2015, 2020, 2025 2030, respectively.

Net Water Demand

To calculate the Net water demand the above-found values considering 30 % deduction were subtracted from the total water supply demand as 8,01,500, 1,58,000, 19,91,500 and 24,50,000 population or the year 2015, 2020, 2025 and 2030 respectively.

Population Projection

$$= \textbf{Net Water Demand x 1000 x 5 (family members)} \quad ...(3.5)$$

The net water demand is multiplied by five (considering a family with five members on average). We got 160.3 MLD, 271.6 MLD, 398.3 MLD, and 490 MLD in 2015, 2020, 2025, and 2030 respectively.

Enroute Village Water Supply

CIDCO provided levels of various nodes of the WTN. 20 MLD of water to enroute villages is supplied through the 35 nodes of the transmission main. On discussion with CIDCO officials, it is decided that this demand of 20 MLD should be equally divided. Hence, the demand of (20/3.5), i.e., 0.5714286 MLD allocation to the primary transmission nodes, has to be made.

In enroute villages, the water demand was 22 MLD, 32 MLD, 39 MLD, and 50 MLD in 2015, 2020, 2025, and 2030.

X-City (Supreme Adhiraj Hex Block Water Demand)

In X-City, the water demand was 00 MLD, 20 MLD, 30 MLD, and 30 MLD in 2015, 2020, 2025, and 2030.

Finally, the total water demand considering all the above factors comes out to be

$$\textbf{Total Water Demand} = (\textbf{Nodes} + \textbf{N} - \textbf{route village} + \textbf{X City})\textbf{demands} \quad \text{...(3.6)}$$

229 + 22 + 00 = 251 MLD in 2015.

388 + 32 + 20 = 440 MLD in 2020.

569 + 39 + 30 = 638 MLD in 2025.

700 + 50 + 30 = 780 MLD in 2030.

Identification of Resources for Current and Future Water Supply Requirements

Table 3.12 show on next page.

Total Water Demand

Patalganga (MJP), Morbe (NMMC), Hetwane (CIDCO), MIDC, Balganga (CIDCO), and Kondhane (CIDCO) are the six reservoirs from where the water resources for supply are considered for the current and future water demands. The exact requirement is listed below in Table no 3.13

Table 3.12: Projected population Vs. Water Supply Demand 2030

Sr.No	Node	Ultimate water demand MLD	Water demand in 2014 in MLD	Water supply consumption 2014 in MLD	Water supply Deficit in MLD	2015	2016	2017	2018	2019	2020	2021	2022	2023	2024	2025	2026	2027	2028	2029	2030
1	Dronagiri	100	12	10	2	13	15	20	25	30	35	40	45	50	55	60	70	80	85	95	100
2	Ulwe	80	7	6	1	8	10	15	20	25	30	35	40	45	50	55	60	65	70	75	80
3	Kharghar & Kharghar hill	130	70	66	4	75	80	85	90	95	100	105	110	105	110	115	120	125	125	127	130
4	Navase - Ph-I / II	25	1	0.5	0.5	2	3	5	7	9	12	15	17	20	22	25	25	25	25	25	25
5	Kamothe	95	40	35	5	45	50	55	60	65	70	75	80	82	84	86	88	90	91	92	95
6	Karanjade	35	1	0	1	3	5	7	10	12	15	18	22	25	27	28	30	32	33	34	35
7	Kalundre	5	1	0.5	0.5	2	3	3	4	4	5	5	5	5	5	5	5	5	5	5	5
8	Talajo	40	4	3	1	5	6	7	10	12	15	18	22	25	27	29	31	33	35	38	40
9	Kalamboli	40	30	27	3	31	2	33	34	35	36	36	37	38	39	39	40	40	40	40	40
10	Panvel	60	44	42	2	45	46	47	48	49	50	52	54	55	56	57	58	59	60	60	60
11	Airport	40	0	0	0	0	0	0	2	5	10	15	30	25	30	35	30	37	38	39	40
12	Pushpak 2.5 FSI	50	0	0	0	0	0	0	3	5	10	15	30	25	30	35	40	63	45	47	50
	Total	**700**	**210**	**190**	**20**	**229**	**220**	**277**	**313**	**346**	**388**	**429**	**492**	**500**	**525**	**569**	**597**	**654**	**652**	**677**	**700**
Deduct losses and commercial consumtion 30 % to derive population				57		68.7	66	83.1	93.9	103.8	116.4	128.7	147.6	150	164.5	170.7	179.1	196.2	195.6	203.1	210
Net Water Demad (NWD)				133		160.3	154	193.9	219.1	242.2	271.6	300.3	344.4	350	374.5	398.3	417.9	457.8	456.4	473.9	490
Population = NWD x 1000 x 5 (family size)				665000		801500	770000	969500	1095500	1211000	1358000	1501500	1722000	1750000	1872500	1991500	2089500	2289000	2282000	2369500	2450000
13	N-route villages in MLD	50	20	20		22	24	26	28	30	32	33	34	35	37	39	40	42	44	47	50
14	X-City (Supreme Adhiraj hex block water demand)	30						5	10	15	20	25	27	30	30	30	30	30	30	30	30
15	**Total water demand**	**780**	**230**	**210**		**251**	**244**	**308**	**351**	**391**	**440**	**487**	**553**	**565**	**602**	**638**	**667**	**726**	**726**	**754**	**780**
				2014		2015	2016	2017	2018	2019	2020	2021	2022	2023	2024	2025	2026	2027	2028	2029	2030
a	Patalganga (MJP)			85			85									85					
b	Morbe (NMMC)			30			30									30					
c	Hetwane (CIDCO)			**150**			150									150					
d	CIDCO (Balganga)						**350**									350					
e	CIDCO (Kondhane)															**250**					
16	**Water Resourses for fulfillment (MLD)**			**265**			**615**									**865**					

Table 3.13: Water Supply distribution from various sources from 2014 to 2030

Year	Patalganga (MJP)	Morbe (NMMC)	Hetwane (CIDCO)	MIDC	CIDCO (Balganga)	CIDCO (Kondhane)	Available water Qty (MLD)	Water Demand (MLD)	Shortfall (+)	Surplus (-)
2014	69	35	86	0	0	0	190	210	-20	
2015	69	35	150	1	0	0	255	251		4
2016	69	35	150	3	0	0	257	244		13
2021	115	35	150	3	350	0	653	487		166
2030	115	35	150	3	350	250	903	780		123

C) Model Generation

The scenario represents a set of models that describes the traits of hydraulic networks of different waterworks. A typical model scenario requires an analysis of several alternatives. Analysis of each alternative requires a separate set of input data explained further and in Table 3. 14.

Table 3.14: Eight alternative cases of Water Transmission Network

Case	Route
Case 1	Existing alignment (Hetawane Dam – Tarankhope – Kopar – Jite)
Case 2	Existing alignment, Work in Progress (Hetawane Dam-Tarankhope-Kopar - Jite)
Case 3	Proposed Alternative Alignment I (along NH-17 via Shitole to WTP, Jite)
Case 4	Proposed Alternative Alignment II Shitole Ambivali via NH-17 to WTP, Jite
Case 5	Future Alignment Alternative 1
Case 6	Future Alignment Alternative 2
Case 7	Future Alignment Alternative 3
Case 8	Future Alignment Alternative 4

Case 1: Existing alignment (Hetawane Dam – Tarankhope – Kopar – Jite)

A 1500 mm dia PSC line of length 11.90 Km. has been laid from the dam to Eirwadi (Tarankhop), and again a PSC pipeline of 1200 mm dia from Tarankhope to Jite (6.90 km) has been laid. The total length is 18.885 km.

Case 2: Existing alignment, Work in Progress (Hetawane Dam – Tarankhope – Kopar – Jite)

A 1500 mm dia PSC pipeline of 10.00 km has been laid up to the Bhogeshwari River, and M.S. of 1500 mm diameter (8.80 km) from Bhogeshwari river to WTP has been laid. The total length is 18.80 km

Case 3: Proposed Alternative Alignment I (along NH-17 via Shitole to WTP, Jite)

In this alignment, 1500 mm diameter- a total of 17.80 km of (M.S. pipeline 6.96 km & PSC 10.416 km) has been considered.

Case 4: Proposed Alternative Alignment II Shitole Ambivali via NH-17 to WTP, Jite

In this alignment, 1500 mm PSC (10 km and M.S. (8.0 km) total 18.00 km length has been considered

Case 5: Future Alignment Alternative 1:

Considering the 2000 mm dia M.S. pipeline (10 km) & 1500 mm dia MS (1.90 km), and 1200 mm dia PSC (6.90 km)

Case 6: Future Alignment Alternative 2:

I am considering the 2000 mm dia M.S. pipe (10.00 km) 1500 mm dia M.S. pipe (8.80 km), a total of 18.80 km, as shown in Figure 3.11.

Case 7: Future Alignment Alternative 3:

Considering the 2000 mm dia M.S. pipe (10.00 km) 1500 mm dia M.S. pipe (8.80 km), a total of 18.80 km

Case 8: Future Alignment Alternative 4:

Considering the 2000 mm dia M.S. pipe (10.00 km) 1500 mm dia M.S. pipe (8.80 km), a total of 18.80 km

Here studying eight alternative alignments and comparing them to give the most feasible transmission line, which would be having the potential to meet the public water demands, as shown in Table 3.13. Water transmission systems may be represented as networks of multiple nodes, e.g., reservoirs, water treatment plants, storage tanks, and hydraulic junctions interconnected by physical links, e.g., pipes. We consider all the factors and design criteria such as pressure, velocity, discharge, friction factor, frictional losses, losses at bends, pipe materials, costs, junctions, different valves. We are also eliminating the surveys required for these kinds of projects with the help of modern software and techniques. The WTN lies in the Raigad district. All the data w.r.t. latitude, longitude, and altitude is collected via a satellite image and mobile application and is processed ahead by importing them to QGIS Software, where the alignment is tagged with their location and attributes. Various pipe properties such as diameter, length, pipe material through referring technique correct them; the file imported to EPANET is also open-source hydraulic software. They are assigned to perform the hydraulic simulation for various causes, as mentioned. Once the analysis is completed, the results are viewed, and a comparative study is done to compare the various cases to select the best alternative alignment for the WTN to be laid.

3.11.2 Selection of Study Area

Hetawane Dam to Jite Water Treatment Plant can be shown in Figure 3.44.

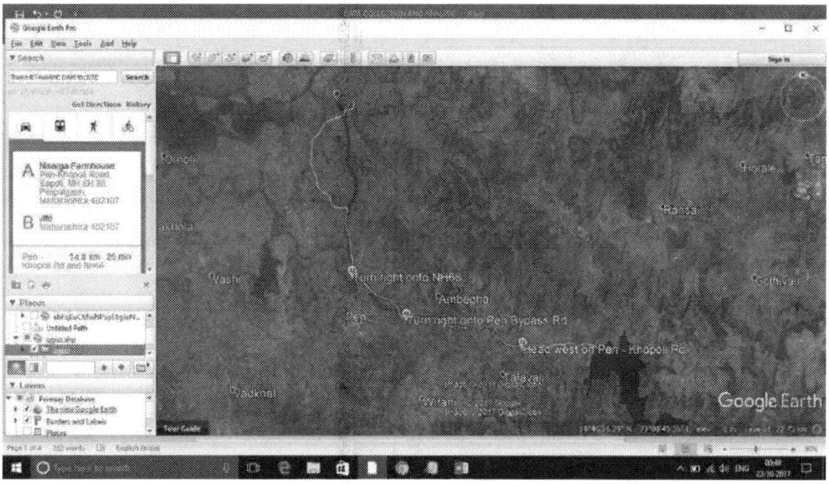

Figure 3.44: Study Area in Google Earth
(**Source:** https://www.google.com/intl/en_in/earth/)

For selecting the study area, we meet CIDCO officials, and after discussing with them, we have selected the area From Hetawane Dam to Jite where the existing WTN is laid. The methodology of WTN is shown in Figure 3.45 on next page.

3.11.3 Collection of Field Data

We have collected the field data with the mobile help app called KOBO collect by creating a server and sending all the data such as latitude, longitude, and altitude. Refer Figure 3.46.

3.11.4 Processing of Collected Data

After opening the kobo collect server and export data in GPS Coordinates (.kml) file format, it is a further process in QGIS software and converted to a (.kml) file. With the help of the export feature being provided in the kobo server, in the downloads section. Refer Figure 3.47.

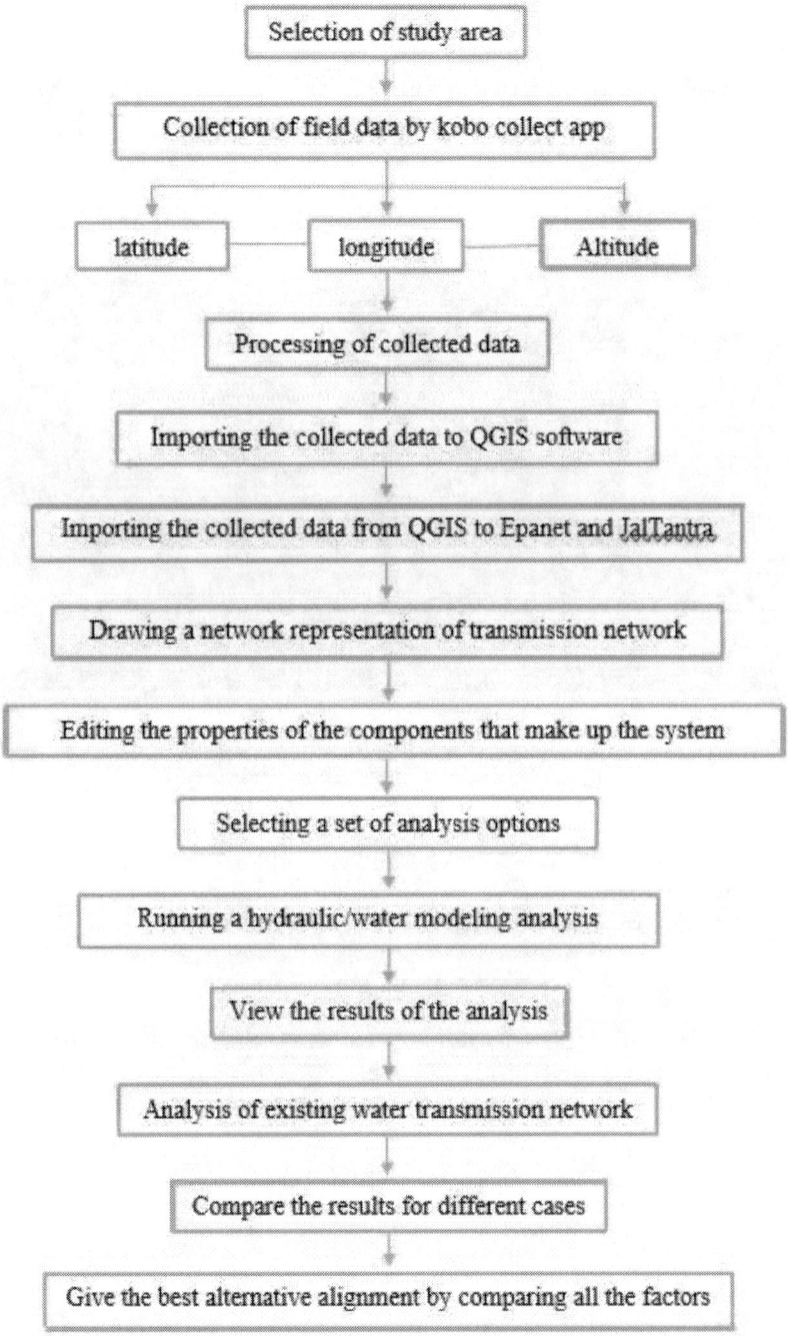

Figure 3.45: Methodology for WTN

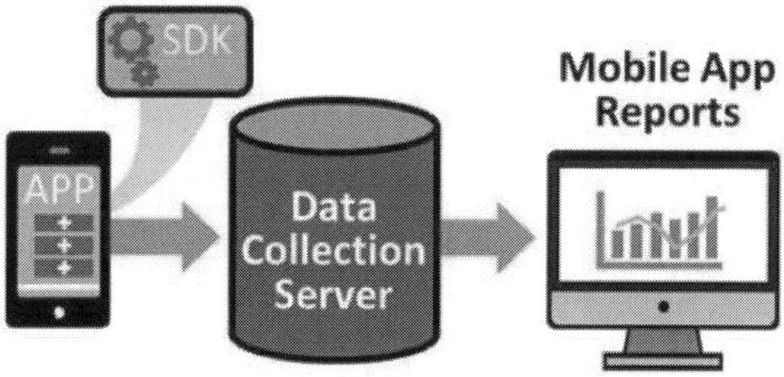

Figure 3.46: Flow Chart - Collection of field data
(**Source:** https://www.google.com)

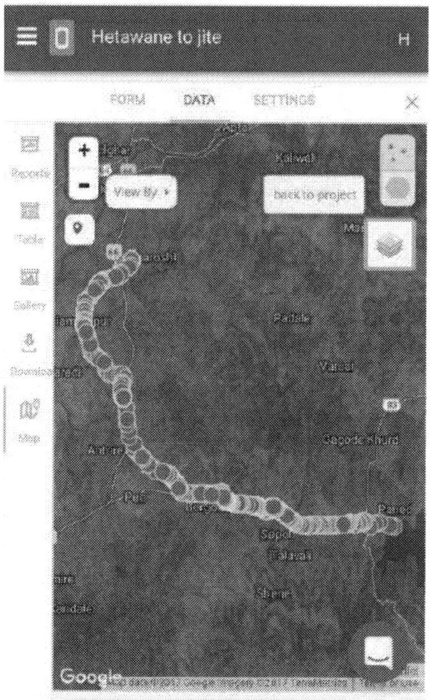

Figure 3.47: Processing of Collected Data, Kobo Toolbox
(**Source:** http://www.kobotoolbox.org/)

3.11.5 Importing the Collected Data to QGIS Software

The imported (.kml) file in QGIS is changed to .shp (shapefile) as it is not possible to import data directly from the kobo collect server to Epanet software; thus, other shapefiles which are needed for pipes, reservoirs, and tanks are also imported to Epanet for further analysis. Refer Figure 3.48.

Figure 3.48: QGIS view (OSV - Open Street Vview)
(**Source:** QGIS Software)

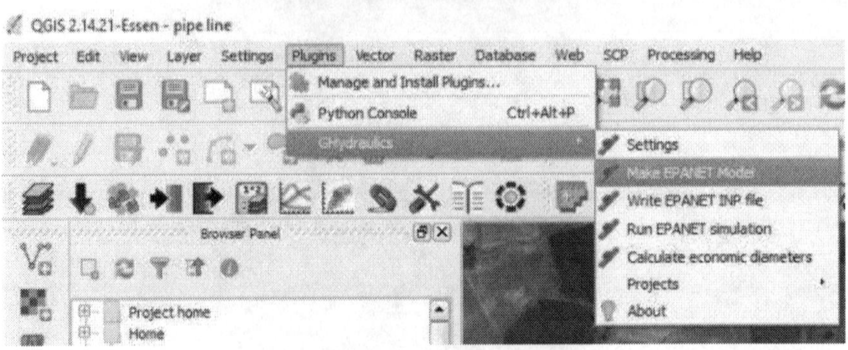

Figure 3.49: G hydraulics Plugin.

A plugin named G-hydraulics is installed in QGIS, wherein all respective layers are transferred in the G-hydraulics plugin. Later converting the file format to .imp should be done by which simulation in Epanet software can be done. Refer Figure 3.49

3.11.6 Importing the Data from QGIS to Epanet

After simulating the hydraulic modeling in QGIS software and making a .imp file, the file is imported into Epanet software by clicking the file menu and then importing the network file to the Epanet software. Refer Figure 3.50.

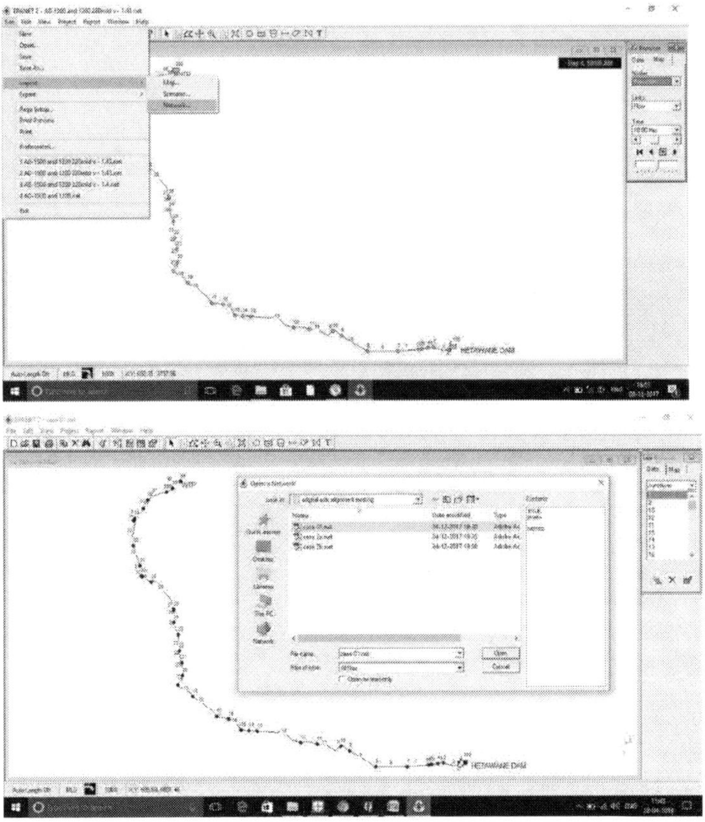

Figure 3.50: Epanet View While Importing Data (Source: QGIS Software)

3.11.7 Drawing a Network Representation of Your Transmission System

By importing the .inp file from QGIS to Epanet network of pipes are drawn. By using various commands in Epanet software such as pipe, we can connect all the nodes with pipes all the nodes of the water transmission network are marked. Provide a reservoir at the start of the network and a tank at the WTP after all the nodes are connected with their respective nodes. Errors can be checked for reservoir during simulation of the network and if any errors are found then those can be rectified. 5.7 editing of the properties of the component that make up the system. Refer Figure 3.51

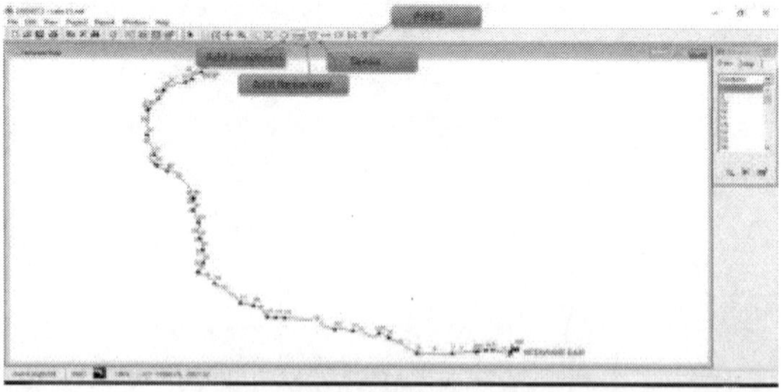

Figure 3.51: Network Representation of Transmission System

After drawing a network, the properties of that network such as diameter, roughness, elevation are assigned. The various pipe diameters used are 1.2 m,1.5 m,2.0 m depending upon the requirement and the topographical condition of the particular area. Pipe materials that have been used are Mild Steel (MS)and Prestressed Concrete (PSC). Based on the material, the roughness coefficient varies. All the elevations of the network are obtaining by using the GIS and RS approach with precise accuracy. Demand of each node is found and for the enroute demand and the final demand is calculated for the WTP. Refer Figure 3.52.

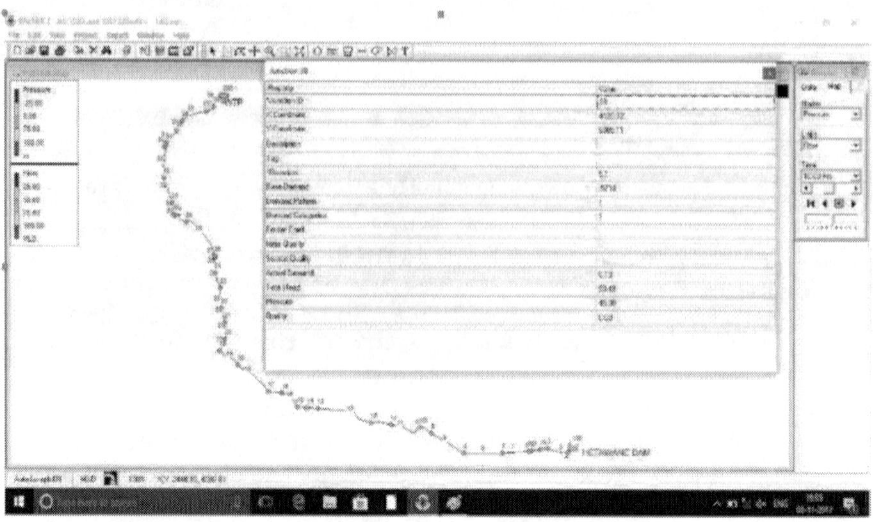

Figure 3.52: Editing Properties of Transmission Network - Epanet

The principal hydraulic input parameters for pipes from the start node and end nodes are the diameter of the pipe, length of pipe, roughness coefficient for determining the head loss, and the status (open, closed, or contains a check valve).

3.11.8 Running a Hydraulic Model

After giving all the inputs and completing the necessary checks, we have to run the hydraulic model, and further, we need to simulate other hydraulics parameters such as pressure, flow, velocity to get the graphical feedback of the output. Refer Figure 3.52 Computed outputs for pipes include flow rate, velocity, head loss and Darcy-Weisbach friction factor. The details information of the WTN layout is shown below in Table 3.15 for the the said eight cases .

Table 3.15: Detailed information of the eight cases of WTN

Case 1: Existing alignment (**Hetawane Dam** to **Tarankhope** to **Kopar** to **Jite**)				
Sr. No	**Phase**	**Diameter (mm)**	**Material**	**Length km**
01	Hetawane Dam to Tarankhope	1500	PSC	11.90
02	Tarankhope to Jite	1200	PSC	06.90
Total Length				18.885 kms
Conclusion: **150 MLD** water reaches the pump house and sump at Jite At **Node J-10,** residual pressure of **2.57 m** is observed.				

Case 2: Existing alignment -Work in Progress (**Hetawane Dam** to **Tarankhope** to **Kopar** to **Jite**)				
Sr. No	**Phase**	**Diameter (mm)**	**Material**	**Length km**
01	Hetawane Dam to Bhogeshwari river	1500	PSC	10.00
02	Bhogeshwari river to Jite	1500	MS	08.80
Total Length				18.80 kms
Possibility A: For Demand =150 MLD, Enroute demand = 20 MLD				
Conclusion: - 150 MLD water reaches the pump house and sump at Jite At Node J-10, residual pressure of 2.57 m is observed.				
Possibility B: For Demand =200 MLD, Enroute demand = 20 MLD				
Conclusion: 200 MLD water reaches the pump house and sump at Jite At Node J-10, residual pressure of 0.31 m is observed.				

<table>
<tr><td colspan="5">Case 3: Proposed Alt. Alignment II (Along with NH-17 vis Shitole Ambivali to WTP Jite)</td></tr>
<tr><th>Sr. No</th><th>Phase</th><th>Diameter (mm)</th><th>Material</th><th>Length km</th></tr>
<tr><td>01</td><td>I</td><td>1500</td><td>MS</td><td>06.96</td></tr>
<tr><td>02</td><td>II</td><td>1500</td><td>PSC</td><td>10.41</td></tr>
<tr><td colspan="4">Total Length</td><td>17.80 kms</td></tr>
<tr><td colspan="5">Possibility C: For Demand =150 MLD, Enroute demand = 20 MLD</td></tr>
<tr><td colspan="5">Conclusion: -
Nodes J-40 to J-46, Negative pressure -13.95m is observed.
A Boring tunnel Machine (TBM) would be required if this alignment is retained for min. 15m depth.</td></tr>
<tr><td colspan="5">Possibility D: For Demand =100 MLD, Enroute demand = 20 MLD</td></tr>
<tr><td colspan="5">Conclusion: -
Nodes J-40 to J-45, Negative pressure -9.29m is observed.
A Boring tunnel Machine (TBM) would be required if this alignment is retained for min. 10m depth.</td></tr>
</table>

<table>
<tr><td colspan="5">Case 4: Proposed Alternative Alignment I
(Along NH-17 vis Shitole to WTP Jite)</td></tr>
<tr><th>Sr. No</th><th>Phase</th><th>Diameter (mm)</th><th>Material</th><th>Length km</th></tr>
<tr><td>01</td><td>I</td><td>1500</td><td>PSC</td><td>10.00</td></tr>
<tr><td>02</td><td>II</td><td>1500</td><td>MS</td><td>08.00</td></tr>
<tr><td colspan="4">Total Length</td><td>18.00 kms</td></tr>
<tr><td colspan="5">Possibility E: For Demand =150 MLD, Enroute demand = 20 MLD</td></tr>
<tr><td colspan="5">Conclusion: -
Nodes J-46 to J-66, Negative pressure -1.45 m min is observed.
Hence pipes can be laid below 2.0 m depth.</td></tr>
<tr><td colspan="5">Possibility F: For Demand =150 MLD, Enroute demand = 20 MLD</td></tr>
<tr><td colspan="5">Conclusion: -
Nodes J-46 and J-66, Negative pressure -1.18 and -1.4 m, respectively, are observed.
Hence pipes can be laid below 2.0 m depth.</td></tr>
</table>

Case 5: Future Alignment (Alternative 1)				
Sr. No	**Phase**	**Diameter (mm)**	**Material**	**Length km**
01	I	2000	MS	10.00
02	II	1500	MS	01.90
03	III	1200	PSC	06.90
Total Length				18.80 kms
Possibility G: For Demand =150 MLD, Enroute demand = 20 MLD				
Conclusion: - 150 MLD water reaches the pump house and sump at Jite				

Case 6: Future Alignment (Alternative 2)				
Sr. No	**Phase**	**Diameter (mm)**	**Material**	**Length km**
01	I	2000	MS	10.00
02	II	1500	MS	08.80
Total Length				18.80 kms
Possibility H: For Demand =**150 MLD**, Enroute demand = **20 MLD**				
Conclusion: - 150 MLD water reaches the pump house and sump at Jite At Node J-10, residual pressure of 5.3 m is observed.				
Possibility I: For Demand =350 MLD, Enroute demand = 20 MLD				
Conclusion: - 350 MLD water reaches the pump house and sump at Jite At Node J-34, residual pressure of 1.35 m is observed.				

Case 7: Future Alignment (Alternative 3)				
Sr. No	**Phase**	**Diameter (mm)**	**Material**	**Length km**
01	I	2000	MS	10.00
02	II	1500	MS	08.80
Total Length				18.80 kms
Possibility J: For Demand =**150 MLD**, Enroute demand = **20 MLD**				
Conclusion: - Nodes J-40 to J-46, Negative pressure -8.34m is observed. A Boring tunnel Machine (TBM) would be required if this alignment is retained for min. 10m depth				

<table>
<tr><td colspan="5">Case 8: Future Alignment (Alternative 4)</td></tr>
<tr><td>Sr. No</td><td>Phase</td><td>Diameter (mm)</td><td>Material</td><td>Length km</td></tr>
<tr><td>01</td><td>I</td><td>2000</td><td>MS</td><td>10.00</td></tr>
<tr><td>02</td><td>II</td><td>1500</td><td>MS</td><td>08.80</td></tr>
<tr><td colspan="4">Total Length</td><td>18.80 kms</td></tr>
<tr><td colspan="5">Possibility K: For Demand =150 MLD, Enroute demand = 20 MLD</td></tr>
<tr><td colspan="5">Conclusion:
150 MLD water reaches the pump house and sump at Jite
At Node J-66, residual pressure of 4.19 m is observed.</td></tr>
<tr><td colspan="5">Possibility L: For Demand =200 MLD, Enroute demand = 20 MLD</td></tr>
<tr><td colspan="5">Refer Table No: 3.22 (Pipe) & 3.23 (Junction)</td></tr>
<tr><td colspan="5">Conclusion:
200 MLD water reaches the pump house and sump at Jite
At Node J-66, residual pressure of 1.59 m is observed.</td></tr>
<tr><td colspan="5">Possibility M: For Demand =200 MLD, Enroute demand = 20 MLD (from dam to WTP at RL 36)</td></tr>
<tr><td colspan="5">Conclusion:
200 MLD water reaches the pump house and sump at Jite
At Node J-66, residual pressure of 1.59 m is observed.</td></tr>
</table>

3.12 WATER DISTRIBUTION NETWORK (WDN)

The nature and scope of this study are to determine different parameters in the water conveyance network of the study area, such as discharge, velocity, pressure, head loss. Which controls the design of pipe, i.e., diameter based on its merits, different pipe materials would be assigned, such as galvanized iron, cast iron, steel, concrete. Moreover, its adherence to the design parameters would be compared.

This study involves locating the most excellent location for placing the water tank over the hilltop with DEM application, as shown in Figure 3.53. They lay the WDN down the hill slope, taking advantage of gravity without utilizing pumps, which corresponded with the study's objective in conjunction with optimization.

The Hazen-Williams method is prevalent, especially among civil engineers, because of its ease of use, and its friction coefficient "C" is not a function of velocity or pipe diameter. Its validity depends on the successful selection of the constant friction loss coefficient "C," which is a function of pipe material, pipe linings, and pipe age to indicate the roughness of a pipe interior. The higher the C factor, the smoother the pipe. The proposed water distribution network would be solved using Hazen William's formulae as it is the best-suited method for solving the gravity-based water distribution networks.

Hydraulic software EPANET is be used for determining discharge, velocity, head loss. For the given WDN. This formula is also quite helpful in comparing various pipe materials in the network as different pipe materials have different roughness factors, which is a defining factor for the Hazen-William formula.

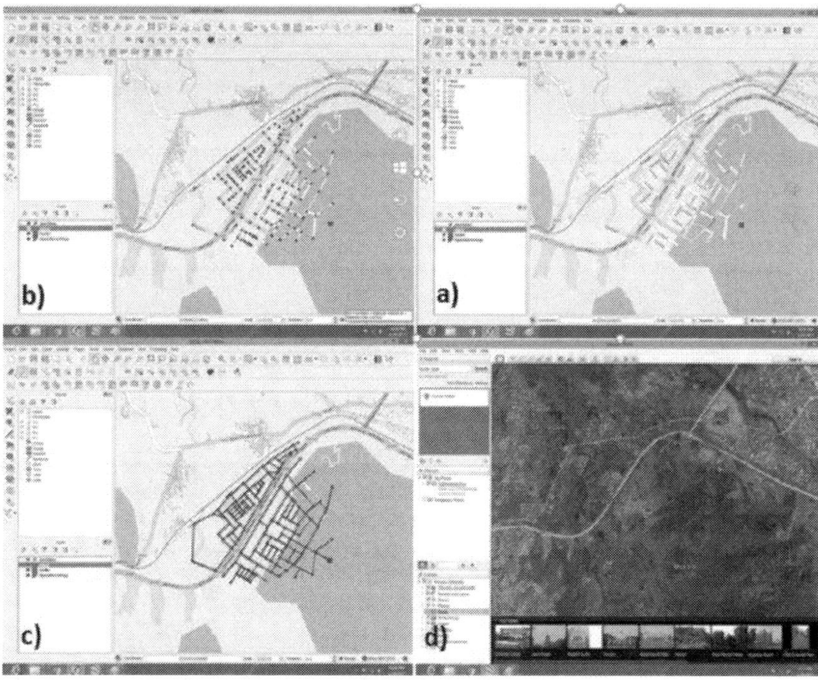

Figure 3.53: Simulation of Water Distribution Network WDN in QGIS

Hazen William Formula is applied to calculate Hf, V, and Q as shown below in equations 3.11,3.1 2 and 3.13, respectively.

$$\mathbf{Hf} = \frac{\mathbf{10.70 \times Q^{1.852}}}{\mathbf{D^{4.87} \times C^{1.852}}} \quad \text{...(3.7)}$$

$$\mathbf{V = 0.354 \times C \times D^{0.63} \times S^{0.54}} \quad \text{...(3.8)}$$

$$\mathbf{Q = 0.278 \times C \times D^{2.63} \times S^{0.54}} \quad \text{...(3.9)}$$

Where,

Q = Discharge, C = Hazen William Coefficient, D = Diameter, S = Hydraulic Gradient L = Length of the pipe, V = Velocity of flow, Hf = Head Loss

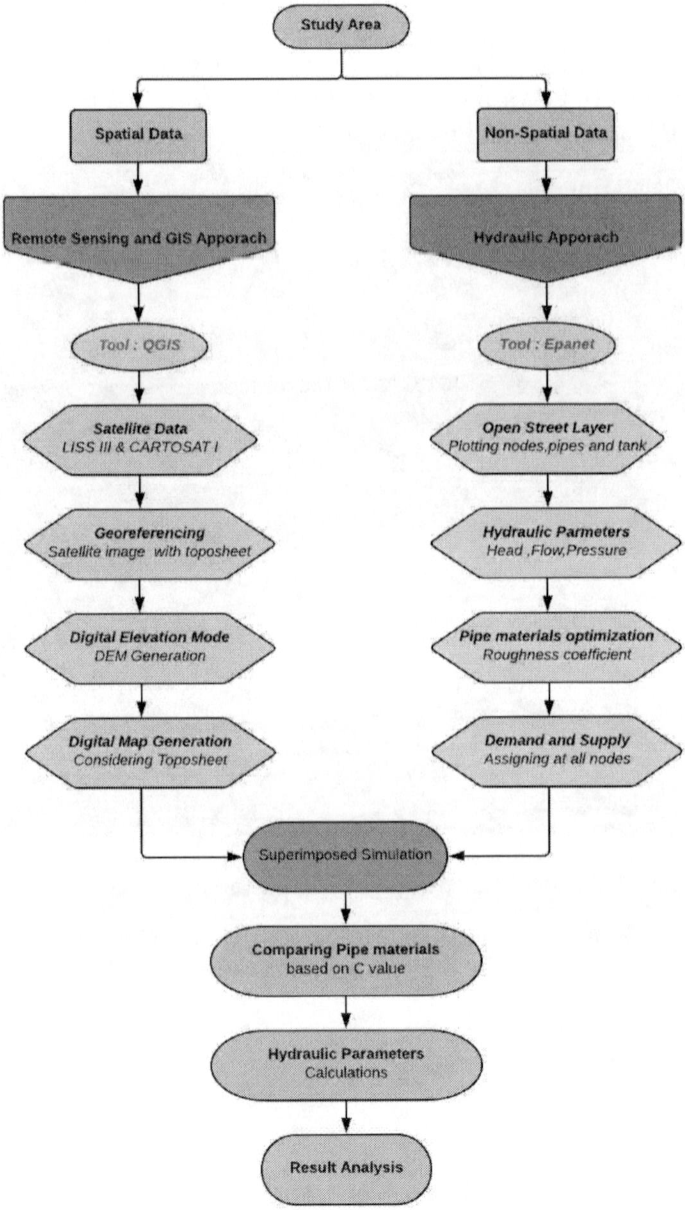

Figure 3.54: Methodology of WDN

Before transferring the network to the simulation, the water tank, pipes, and nodes were plotted in QGIS with Open Street map Plugins. The WDN is imported from QGIS to EPANET, as shown in Figure 3.55. A hydraulic model of the study area is set up for simulation in EPANET. All the nodes and the pipes are installed in the desired location with QGIS and satellite data. As explained earlier, and also the demand for the given water system is taken into account.

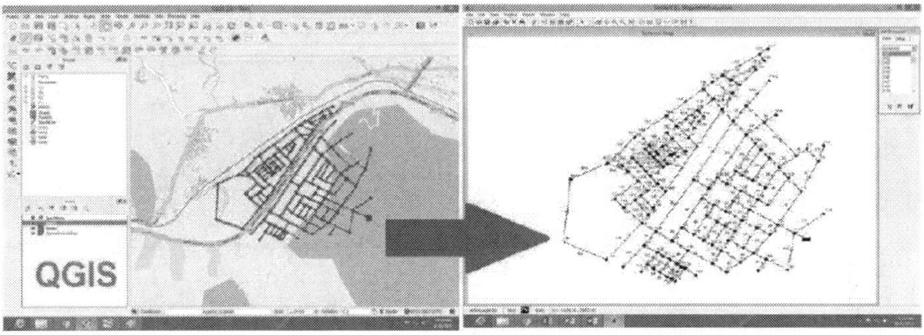

Figure 3.55: Shifting of Networking from QGIS to EPANET

For the node's parameters, elevation DEM provided its value and demand surveys, and the studies were conducted for the given water distribution system. For the pipes, data such as length were calculated by the Open layer, a QGIS Plugins and diameter is defined as suiTable for the proposed system itself. Reservoirs or tanks are set up in the hillock, which as per DEM, is the highest elevation in that region to avail gravity flow, and boundary conditions for them are also being defined.

3.12.1 Hydraulic Simulation

The Hazen-Williams equation is used widely in water supply and sanitary engineering. This equation uses a constant, the Hazen-Williams C, to indicate the roughness of a pipe interior. Because of the empirical nature of the equation, its range of applicability is limited. Many textbooks and software manuals give C values based on pipe type, condition, and age but do not give the range of applicability. Historic experimental data is used to demonstrate that C is a vital function of Reynolds number and pipe size and that the Hazen-Williams equation has narrow applicable ranges for Reynolds numbers and pipe sizes.

The level of error when the Hazen Williams equation is used outside its data ranges is significant. However, a valid C for a given pipe at a specific Reynolds number can be used to estimate a pipe's relative roughness, which then can be used by the rational Darcy-Weisbach equation without the range limitations.

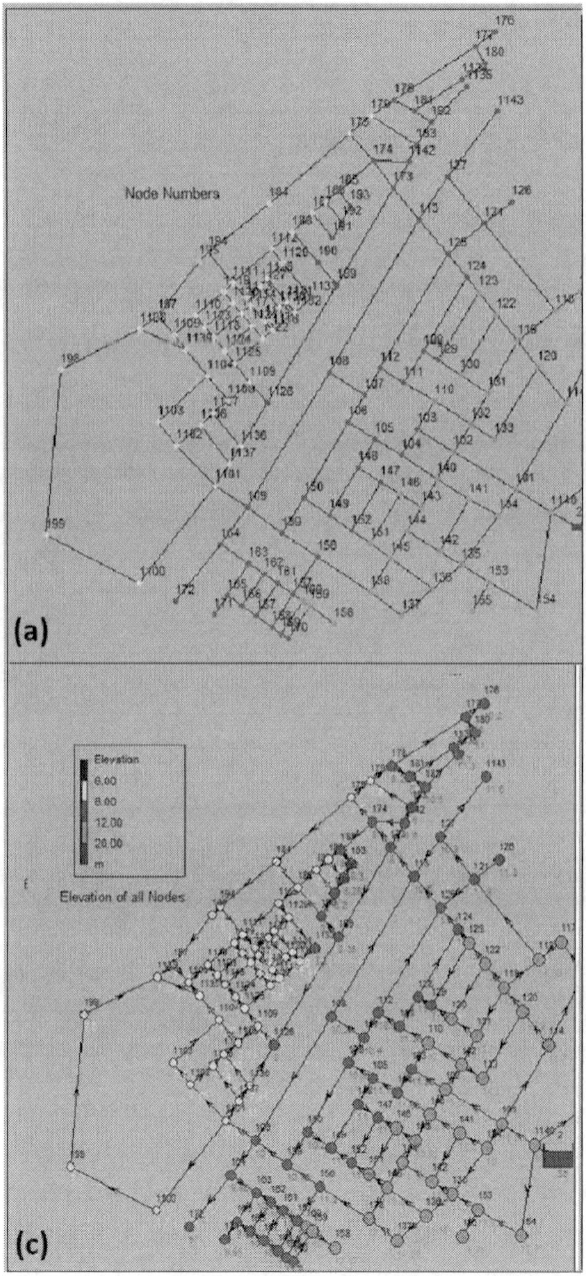

Figure 3.56: (a) Nodes (junctions) numbering (c) Elevation of all the nodes

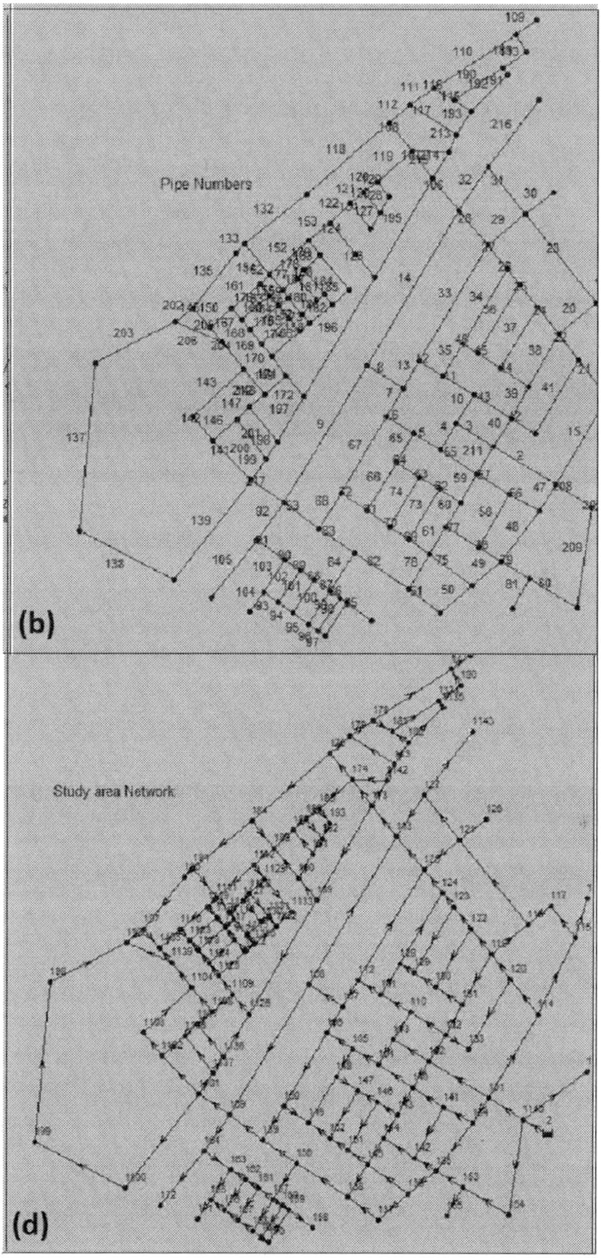

Figure 3.57: (b) Links numbering (d) Flow directions due to gravity.

In a single period, analysis, the network was set up, and then it is allowed to run for that instant only. After that, the system checks all the parameters and determines whether it would be able to meet the water demands or not. Figure 3.56 and 3.57 shows the entire WDN constructed in EPANET to simulate the study area. In all, there are 142 nodes or junctions and 207 pipes or links which are assigned with numbers. These numbers assigned help to know in identifying the various nodal or link parameters as and when required.

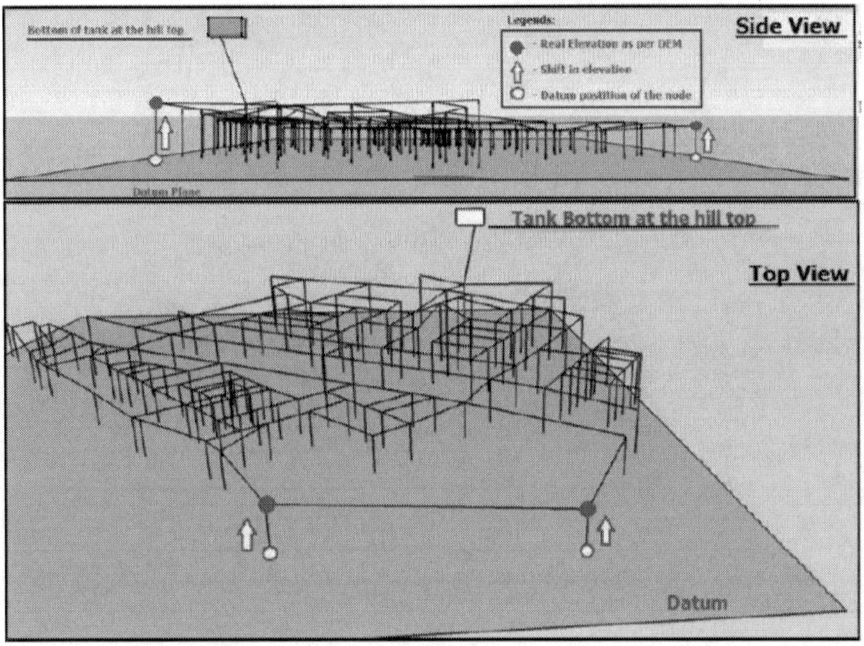

Figure 3.58: 3D view of the WDN: elevation of nodes coinciding as per DEM

The elevation found out by DEM through GIS application played a vital role here, i.e., these elevations were compared with the toposheet and then assigned to all the 142 nodes without any conduction of field survey. The water tank is decided to be placed at the top of the nearby hill, which has its tank bottom at 35.00 m R.L so that it can supply the water to the study area through WDN at the lower level of its foothill. In Figure 3.58, the circular symbols show the node position such that the bigger the node size higher the elevation, and vice versa. To visualize the elevation difference between the tank and the other nodes horizontal plane at 0.00 m is plotted. Moreover, concerning that plane, all the other nodes and the tank are allotted their elevation. A 3D visualization is made in sketch-up software, which can be seen in Figure 3.58.

3.12.2 Simulation of Water Distribution Network

The conclusion is described in graphical and tabular formats, stating that velocity varies from 2 to 0.7 m/s. The head varies from 25 to 6 m, and flow varies from 34.7 to 1.93 MLD within its limits (refer to Figure 3.59). The significance of the results has its direct implications as well as can be adopted for practical applications of laying the WDN where the geography of the terrain can be appropriately identified through remote sensing and GIS via satellite data, and an optimized network can be placed at a developed site which is under rehabilitation. To sum up, a demanding water area can be provided with a 24 x 7 water supply by applying the latest GIS technologies by which above-ground supply can be provided through Elevated Service Reservoir (ESR) where DEM can be used for finalizing its location.The developed simulated models were tested for un observed inputs mainly with varying diameters of pipes and pipe materials, and their qualitative and quantitative performance was judged utilizing demand and supply between the observed and forecast values for a single period and extended period simulation with Hazen Williams equation with DEM generated data for the elevation of nodes. In addition, a radar diagram was plotted, as shown in Figure 4.30, to visualize the nature of pipe diameters utilized overall in the network to understand the percentage of optimization that can be achieved with various trials. These pipe diameter trials also justify the type of pipe material; it is roughness coefficient based on internal smooth and rough pipe wall boundaries. Elevations of all the nodes, and diameters can be seen in Figure 3.60.

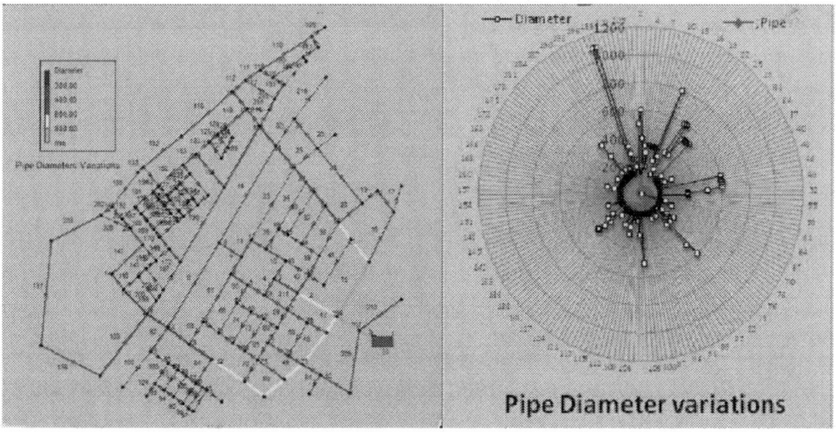

Figure 3.59: Varying diameters of pipe from the source to the destination and Radar diagram of diameter values to a center point for pipe optimization studies

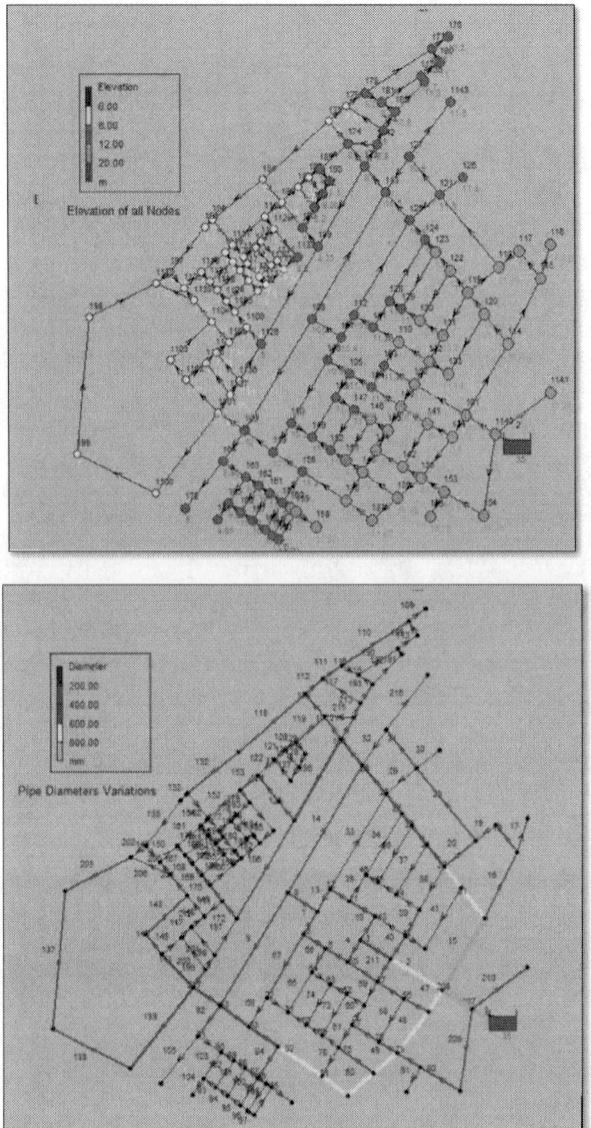

Figure 3.60: Elevations of all the nodes (bigger the nodes, higher the elevation, and vice versa) and diameters of pipes plotted by DEM

3.12.2.1 Comparison of Pipe Materials

Various pipe materials having a different value of roughness constant were modeled and tested through the simulation, and the material which is best-suited corresponding to it pipes boundaries roughness coefficients were found out. A random 13 pipes head loss results are shown in Table 3.16 for iron, concrete, fiber, and PVC pipes.

Table 3.16: Roughness coefficient C values of various pipe materials.

IRON	(C=100)	CONCRTE	(C=120)	FIBER	(C=140)	PVC	(C=150)
	Headloss		Headloss		Headloss		Headloss
Link ID	m/km	Link ID	m/km	Link ID	m/km	Link ID	m/km
Pipe 2	4.68	Pipe 2	3.34	Pipe 2	2.51	Pipe 2	2.21
Pipe 3	10.72	Pipe 3	7.65	Pipe 3	5.75	Pipe 3	5.06
Pipe 4	14.33	Pipe 4	10.22	Pipe 4	7.68	Pipe 4	6.76
Pipe 5	8.71	Pipe 5	6.22	Pipe 5	4.67	Pipe 5	4.11
Pipe 6	9.19	Pipe 6	6.55	Pipe 6	4.93	Pipe 6	4.34
Pipe 7	7.03	Pipe 7	5.02	Pipe 7	3.77	Pipe 7	3.32
Pipe 8	7.89	Pipe 8	5.63	Pipe 8	4.23	Pipe 8	3.72
Pipe 9	3.36	Pipe 9	2.4	Pipe 9	1.8	Pipe 9	1.58
Pipe 10	12.29	Pipe 10	8.77	Pipe 10	6.59	Pipe 10	5.8
Pipe 11	12.55	Pipe 11	8.95	Pipe 11	6.73	Pipe 11	5.92
Pipe 12	8.59	Pipe 12	6.13	Pipe 12	4.61	Pipe 12	4.06
Pipe 13	5.03	Pipe 13	3.59	Pipe 13	2.7	Pipe 13	2.38
Pipe 14	1.57	Pipe 14	1.12	Pipe 14	0.84	Pipe 14	0.74

3.12.2.2 Profile Plots

It is observed that the higher the C value of pipe, i.e., pipe with smooth surfaces has a lesser head loss and vice versa. The simulation model for single-period analysis exhibited good performance in testing where the flow in a pipe directly proportional to its demand, showing variations from 0.26 mld to 3.53 mld per day, as shown in Figure 3.61. The plot flow in the ordinate axis and pipe or link number in the abscissa axis confirm. It can be seen that the majority of the flow is positive. As per the continuity equation and basics, hydraulic principles flow in a pipe is directly proportional to its demand and its utility. Hence, to make the WDN more sustainable, water in mld is calculated, and its simulated output all over the network is shown in Figure Figure 3.61, where flow variations from 0.26 mld to 3.53 mld per day is shown.. Figure 3.62 shows the implication of elevation of all the nodes assigned in the WDN. These nodes have a direct impact on the unit head loss. It can be seen that the elevation varies from 6 m minimum to 35 m maximum where the water tank is placed. As the elevation is directly proportional to the head loss, its variation can be seen to vary from 0.40 m / km to 1 m /km. Figure 3.63 shows the elevation (m) plotted with DEM and its effects on velocity in a pipe in m/sec. The plot shows pipe number Vs. velocity, which directly reflected the ground slope and its variation in velocity.

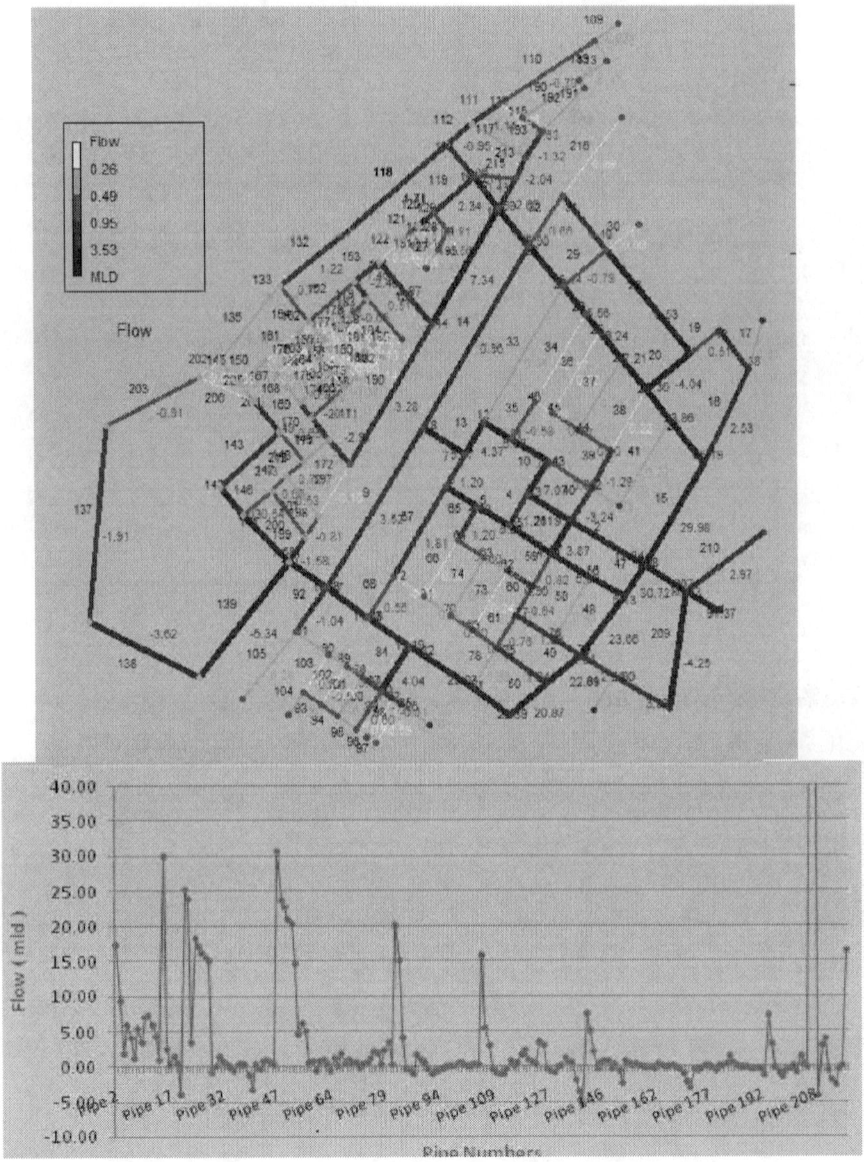

Figure 3.61: Flow variations in the pipes, which ranges from 12 mld to 40 mld

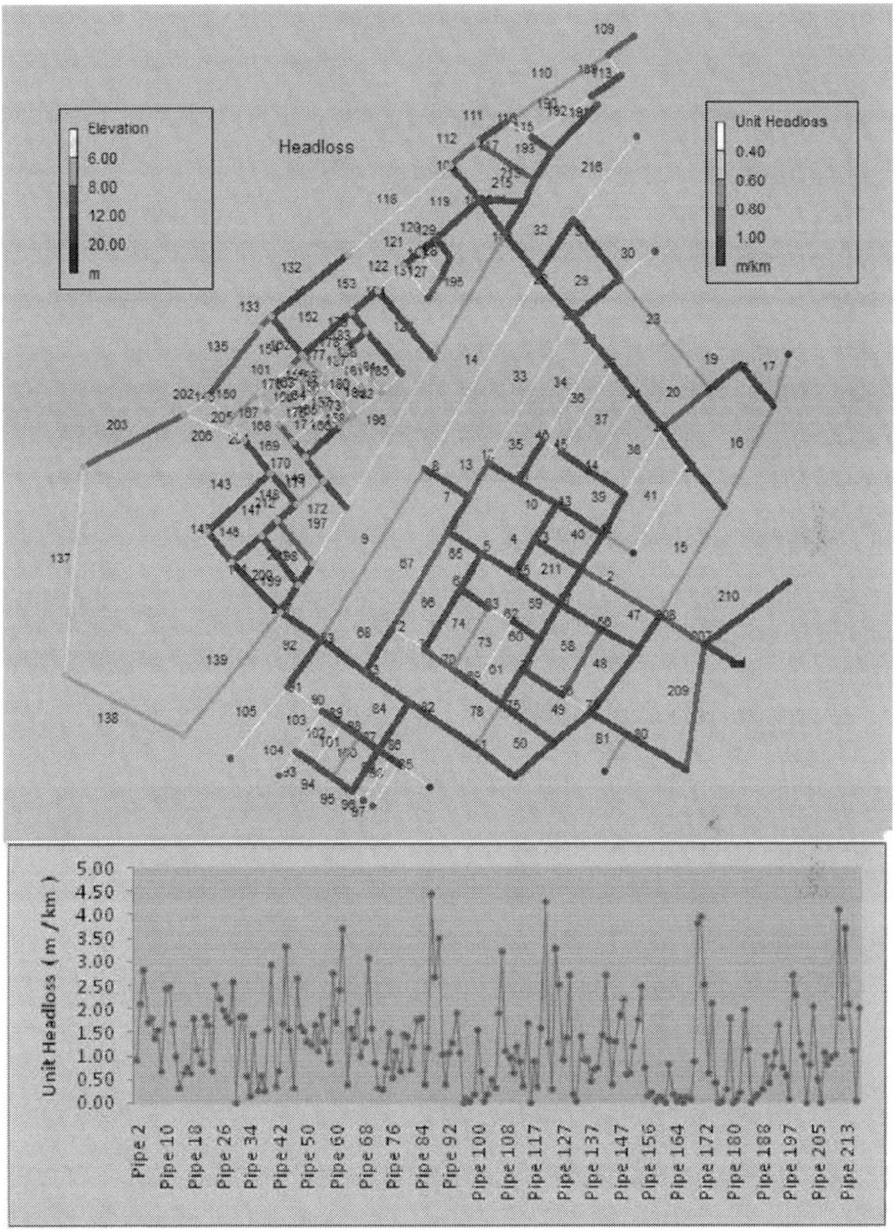

Figure 3.62: Elevation (m) and its effects on the head loss (m/km)

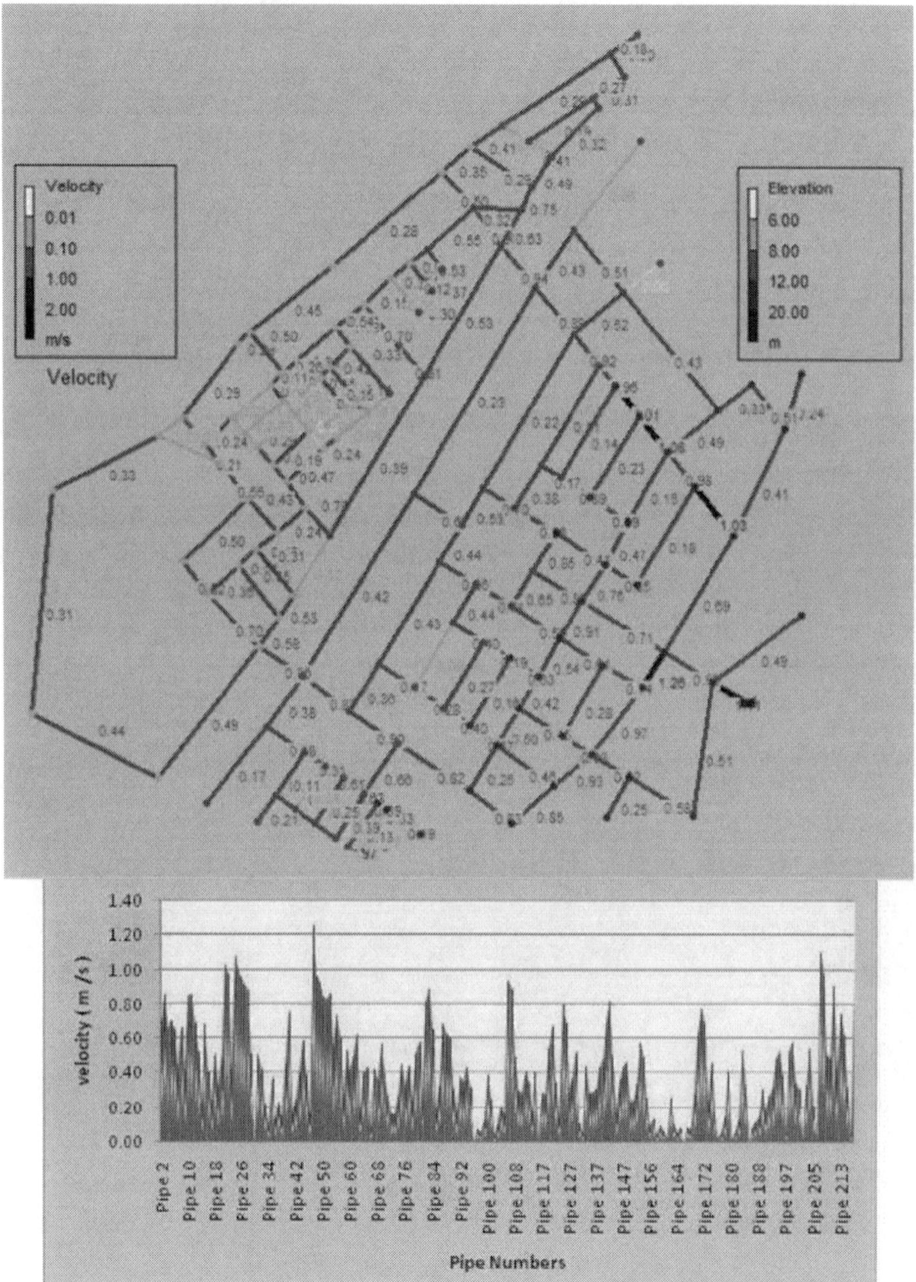

Figure 3.63: Elevation (m) & its effects in the velocity (m/sec)

Network Results

In a single period, analysis, the network is set up, and then it is running for that instant only. After that, the system checks all the parameters and determines whether the system would be able to meet the water demands or not.

A demand pattern (Refer Fig 3.64) represents a factor by which the demand would be multiplied depending upon the time of the day. Generally, a 24-hour pattern is set at a one-hour time step. Following is the demand pattern which is being used for the given network. After the demand pattern is set up, the system is running again, and results are obtained.

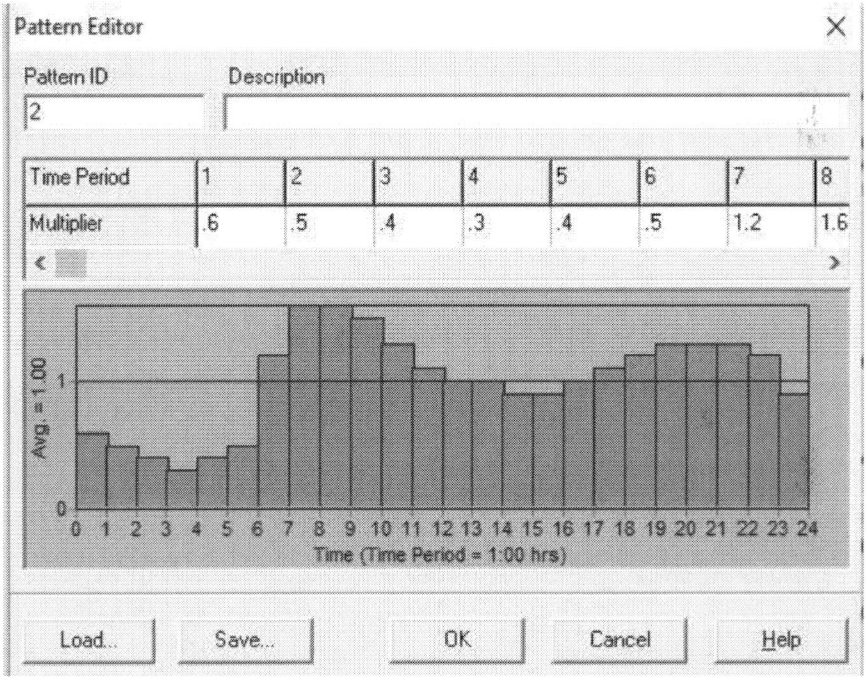

Figure 3.64: Hour demand pattern input in EPANET

Figure 3.65 shows a symmetrical variation that can be seen in Time Series Plot, which shows three significant parameters such as pressure, demand, and head value versus time, where pressure and head are almost parallel to each other, and the demand is precisely its reverse. Results show that as the demand increases, supply decreases, and hence their effect on pressure and head dips down.

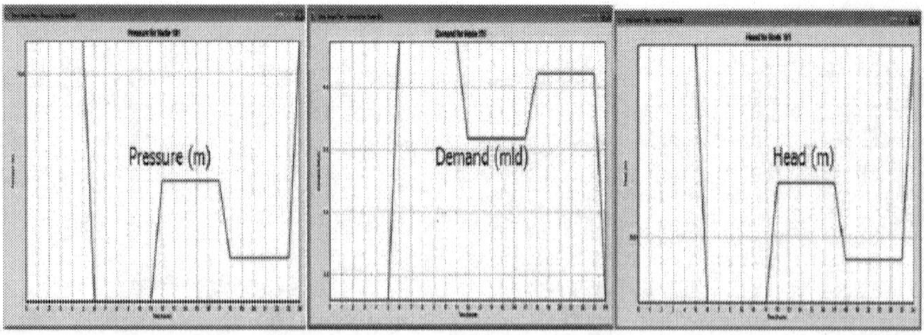

Figure 3.65: Time Series Plot of various parameters such as Pressure (m), Demand (mld), and Head (m) in the ordinate axis and time (24 hrs) in abscissa axis for Node 10 of the WDN

Figure 3.65 shows a Profile Plot of demand Vs. all network junction at 0:00, 08:00, 16:00, and 23:00 hours respectively. Again, an apparent dip in demand can be seen as the duration of the day moves ahead.

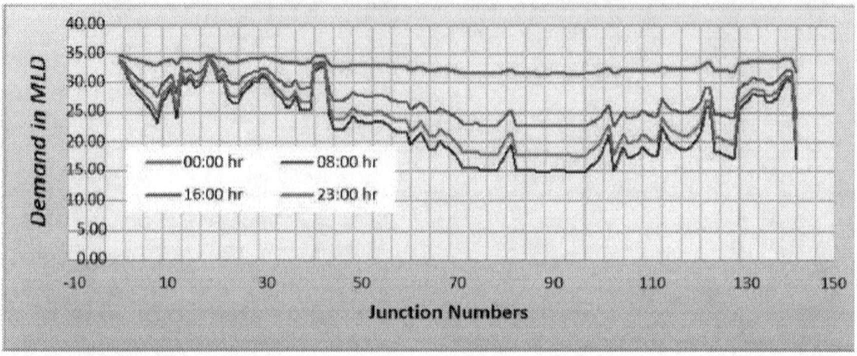

Figure 3.66: Graphical presentation of varying demand Vs. all junctions in the network.

Figure 3.67, which shows contour plots of pressure, demand, and head, justifies the statements and plots shown in Figures 3.66 and 3.67.

3.13 WATER QUALITY

In the water distribution network (WDN), the secondary disinfection method is the only way to control the water quality, whereas before it, we have primary disinfection and after it tertiary disinfection. However, tertiary disinfection (UV disinfection, Reverse Osmosis) is one of the most expensive methods to control water quality according to consumers' needs. In contrast, the problem is that only some people can afford to install this tertiary disinfection method at their houses

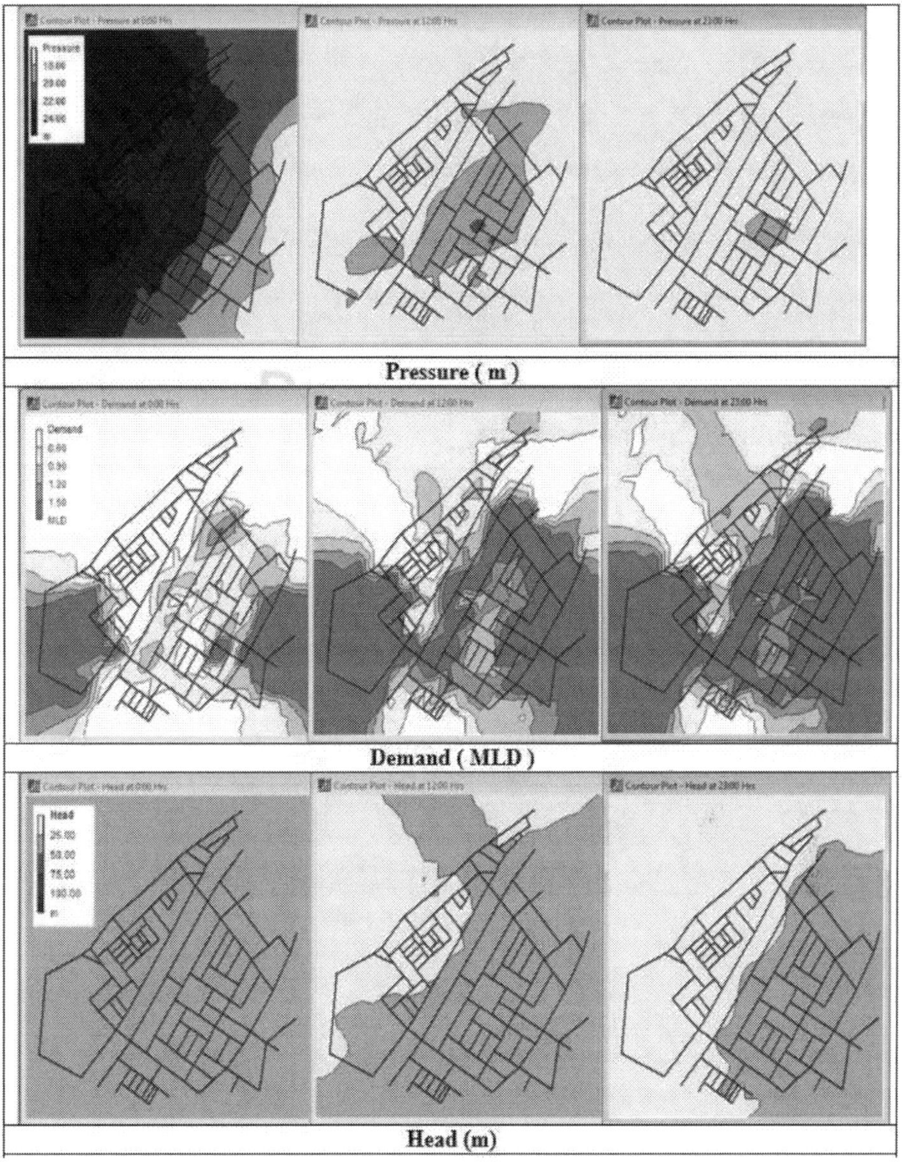

Figure 3.67: Contour Plot of Pressure, demand, and head at 00:00 hrs, noon hrs, and 23:00 hrs respectively of the whole study area

and offices. Thus, to control water quality in WDN, we have chosen the secondary disinfection method because there may be less use of tertiary disinfection.

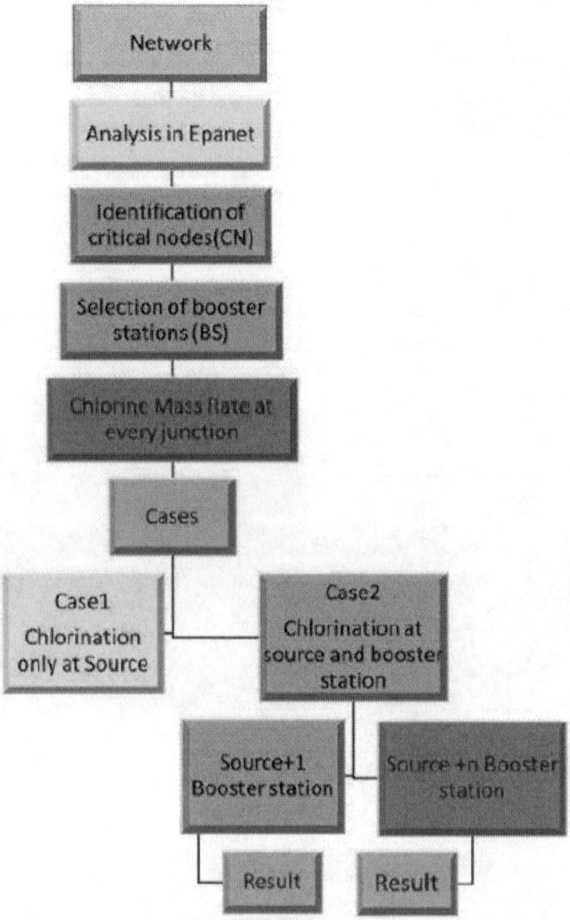

Figure 3.68: Methodology of Booster Chlorination Technique

The study's objective is to create a contaminant-free distribution network with the help of chlorine booster stations, simulation software. Moreover, to work out the optimum location of booster chlorination stations, the technique is shown above in Figure 3.68. Hydraulic and water quality analysis is necessary to predict the outcome. The analysis can predict contaminant spatial and temporal variations throughout the distribution network. The steady-state simulation tool, EPANET, is used as a hydraulic and water quality simulator in pressurized pipe networks. The model can predict flow and pressure at various points in a distribution network using an extended period simulation from one day to one week.

In secondary disinfection, the best parameter available to efficiently control the water quality is residual chlorine levels in WDN. Chlorine has the property to decay itself concerning time. Thus, to control the chlorine levels in WDN, various methods have been used in this methodology. Refer Figure 3.69.

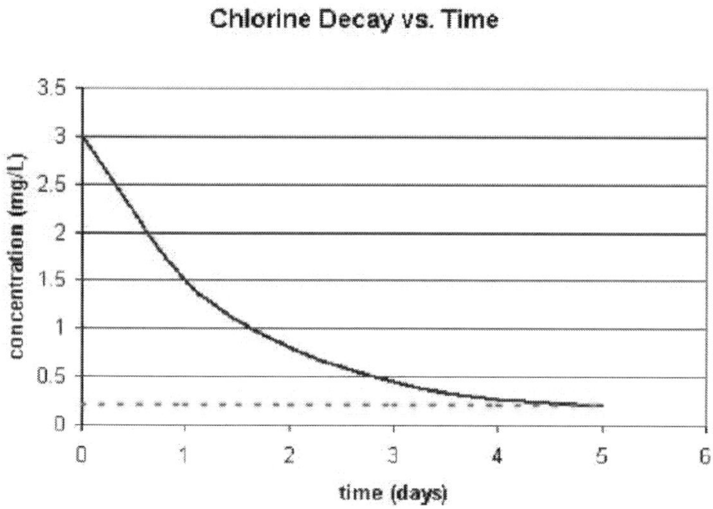

Figure 3.69: Chlorine Decay graph w.r.t time.

3.13.1 Data Required

Physical data: Flow conditions, temperature, pipe material, length of pipe, diameter, base demand & elevation at each node, the population density of area served, the value of friction factor, the velocity of water in the pipe.

Chemical and biological data: Initial chlorine concentration totals organic content, values of decay coefficients such as chlorine decay and wall decay. The total demand for the network is 74.06 MLD. The value of kb is kept as 0.55 per day, and the initial chlorine concentration is 0.5mg/l. Moreover, a demand pattern is used in the network to decide the peak time at which it is maximum.

3.13.2 Analysis in Epanet

- The required distribution network is copied to EPANET for analysis.
- All input data is entered into the software.

- An extended period simulation is carried out for chlorine in water quality in EPANET to the critical nodes with residual chlorine less than 0.5mg/L.

3.13.3 Optimization

At selected nodes, chlorine boosters are provided, and hydraulic simulation is done to determine the amount of chlorine before and after the booster station. Check for an effective booster station by running EPANET software separately for each booster station, and collect the results after each run which shows the effect on all critical nodes from that booster station. From all results, we can make one single Table that can effectively show the effect on all critical nodes due to all booster stations, respectively.Figure 3.70 shown a model of booster station. Calculate the rate of chlorine that is to be injected from every booster station and the source reservoir. Base demand at every booster station node and reservoir node and the amount of chlorine that is to be injected per liter of water, thus multiplying both of them with each other separately for each booster station node and reservoir node, we can calculate the rate of chlorine mass to be injected totally into the network.

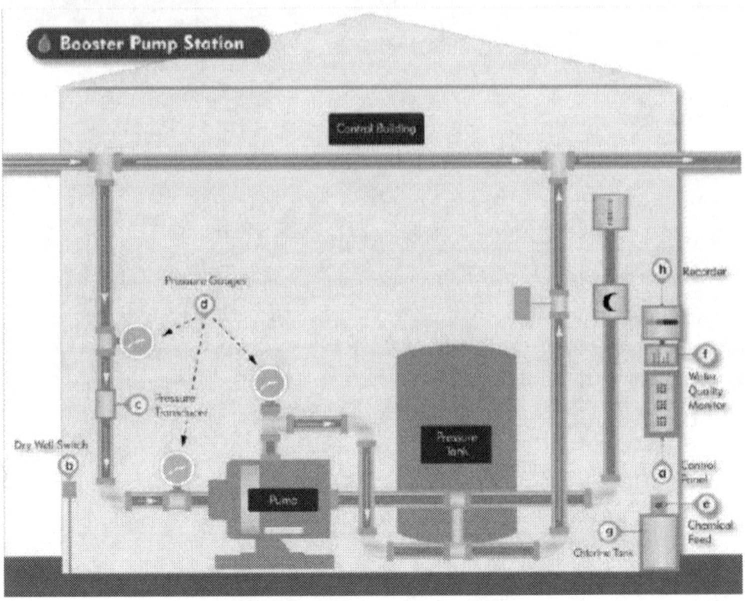

Figure 3.70: Chlorine booster station
(**Source:** Geoscience Journal)

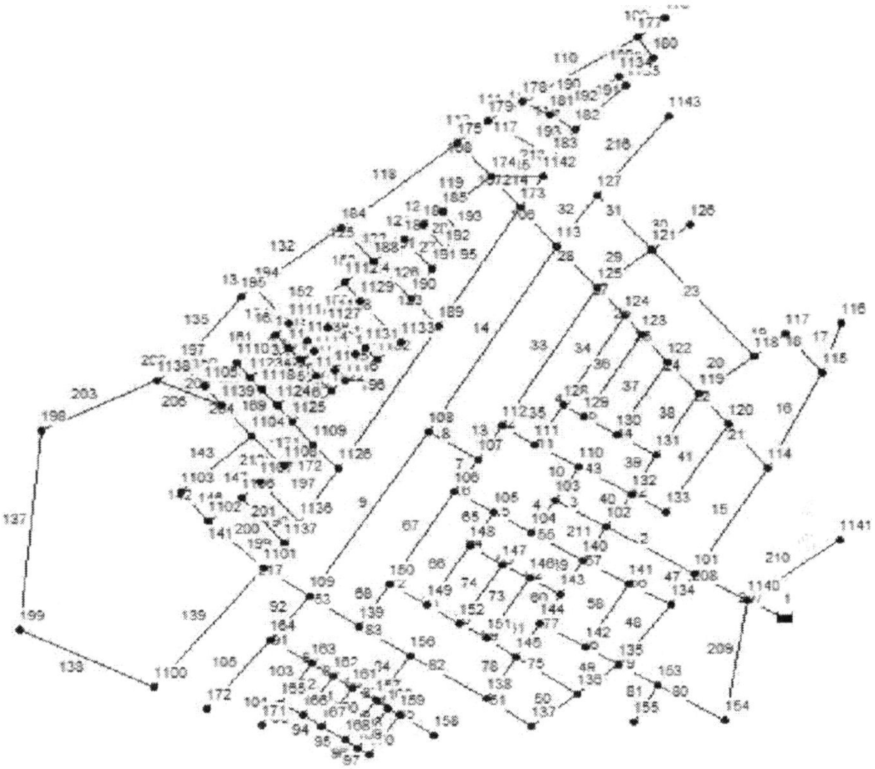

Figure 3.71: General Layout of the network with node ids and pipe ids.

3.13.4 Residual Chlorine Simulation in Water Quality

Water distribution network is primarily designed in EPANET using the data and is run for hydraulic simulation analysis in EPANET. Maintaining residual chlorine as 0.5mg/l as per IS 10500 (2012), the initial quality of water is set in the software, and it is run for 2 hrs. (Intermittent water supply). As per the first-order decay kinetics of chlorine, it tends to decay concerning time, so if we maintain the 0.5mg/l chlorine at the end of the water treatment plant, then also after some time, the chlorine is going to drop by reacting with organic matter, pathogens, bacteria, and virus. The points where the chlorine levels are less than the prescribed amount are identified as the critical nodes (CN). Refer to Table 3.18.

Table 3.18: Amount of chlorine at different critical nodes (RS1).

Sr. No.	Junction	Chlorine (mg/Lit)
1	Junction 1105	0.16
2	Junction 107	0.16
3	Junction 1138	0.1
4	Junction 1110	0.1
5	Junction 1123	0.12
6	Junction 1120	0.17
7	Junction 1121	0.17
8	Junction 1116	0.17
9	Junction 1115	0.08
10	Junction 1114	0.13
11	Junction 1130	0.13

The result obtained from the network simulation shows the chlorine levels available in WDN after 2-hours of simulation. As the water reaches the end part of the network, the chlorine levels drop drastically and show a color change in the network during analysis. From the results, the critical nodes are identified as nodes where chlorine levels are lesser than 0.2mg/l.

These critical nodes (refer Figure 3.72) are then used to find out the locations of chlorine booster stations in the network. From the results, it can be easily seen that the chlorine levels in the network are good enough in an area closer to the source but the area away from the source the inadequate levels of chlorine which are not suiTable for consumption of that water by consumers.

Boosters are placed at different critical nodes with different chlorine mass rates to injected, and the network is run to come up with adequate chlorine results in WDN by trial and error method. Finally, the best combination of booster stations is selected, and the chlorine mass rate to be injected from each booster is fixed. The result obtained after this is termed RS2. The total demand for the network is 74.06 MLD. The value of kb is kept as 0.55 per day, and the initial chlorine concentration is 0.5mg/l.

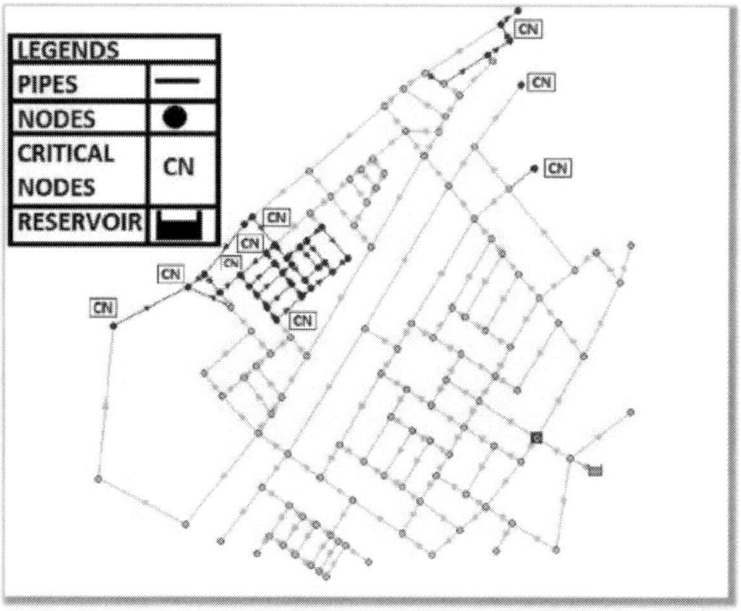

Figure 3.72: Network Showing the CRITICAL NODES (CN)

During the analysis, different materials for pipe (steel, concrete, PSC, plastic) were tested, but no significant change was observed in the decay pattern of chlorine. It was done by making changes in the roughness coefficient. By hydraulic simulation software, extended simulation is done for 2 hrs, and the critical points are identified. At selected points, booster stations are provided such that a minimum amount of residual chlorine is maintained in the network. The following formula determines the chlorine mass rate be injected at each node:

$$\boldsymbol{Mi}\ \left(\frac{\text{mg}}{\text{min}}\right) = \mathbf{Ci}\ \left(\frac{\text{mg}}{\text{l}}\right) \text{x}\ \mathbf{Q}\ \left(\frac{\text{l}}{\text{min}}\right) \quad \ldots(3.10)$$

Where:

Mi = Chlorine mass rate in (mg/min) Ci = Constraints

Q = Discharge (l/min)

The equation used for optimization and calculation is: Minimize:

$$\sum_{i=1}^{n} M_i$$

Subject to constraints:

$$C_j = \sum_{j=1}^{m}\sum_{i=1}^{n} K_{ij}M_i \geq 0.2 \qquad \text{...(3.11)}$$

Table 3.19: Chlorine mass rate Injection.

Injection node	Mi Chlorine mass rate to be injected at each node (mg/min)
M1 Node 198	277.76
M2 Node 1139	312.48
M3 Node 1125	94.4384
M4 Node1123	23.6096
M5 Node 1110	79.856
M6 Node 194	0
M7 Node 1115	27.776
M8 Node 1112	94.7488
M9 Node 127	144.4352
M10 Node 178	63.8848
M11 Node 1135	16.6656
M0 Reservoir	25715.28

Non-negativity constraints,

$M_i \geq 0.$

i = Injection nodes

j = critical nodes

m = total number of critical nodes.

n = total number of injection nodes.

C_j =chlorine concentration at junction nodes 01 to 11 corresponding to critical nodes, respectively, mg/l.

K_{ij} = constants corresponding to injection nodes and critical nodes.

M_i = mass rate injected at injection node corresponding to injection location at source and booster station, mg/min. Refer Table 3.19

Solving the above equation, we get various combinations of booster stations along with the source chlorination.

Further, two cases are considered for optimization of the network. In case 1, the only source is considered for chlorine injection. In case 2 one booster station with source and two booster station is considered. Refer Table 3.20.

Table 3.20: Different Booster Combinations.

Sr. No.	Different combinations	Chlorine mass rate (mg/min)
1	M0+M1	2571.278
2	M0+M2	25715.278
3	M0+M3	25691.668
4	M0+M4	25679.863
5	M0+M5	25715.278
6	M0+M6	25715.278
7	M0+M7	25715.278
8	M0+M8	25691.591
9	M0+M9	25679.169
10	M0+M10	25699.306
11	M0+M11	25711.11

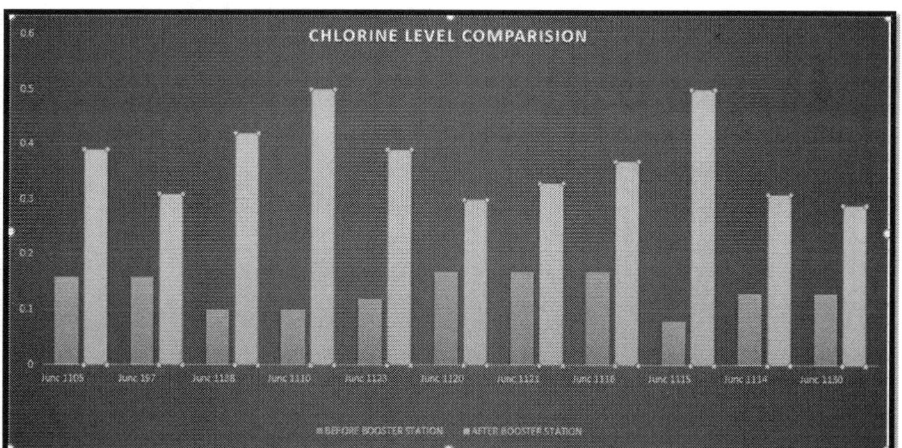

Figure 3.73: Amount of chlorine before and after booster station

Table 3.21: The amount of residual chlorine before and after booster station.

Node	Chlorine before BS	Chlorine after BS	Node	Chlorine before BS	Chlorine after BS
ID	Mg/L	Mg/L	ID	Mg/L	Mg/L
Junc1105	0.16	0.39	Junc1120	0.17	0.3
Junc197	0.16	0.31	Junc1121	0.17	0.33
Junc1138	0.1	0.42	Junc1116	0.17	0.37
Junc1110	0.12	0.5	Junc1115	0.08	0.5
Junc1123	0.17	0.39	Junc1114	0.13	0.31

After analyzing, critical nodes were identified, and booster stations were provided. Finally, the amount of chlorine mass rate to be injected (refer Figure 3.73) is determined and optimized. The rate of chlorine decay in water does not depend much on the material of the pipe used. The amount of residual chlorine before and after booster station is shown in Table 3.21.

Table 3.22: Chlorine booster at Reservoir.

JUNCTION NO.	BASE DEMAND (MLD)	CHLORINE Cj (mg/l)
1105	0.18	0.16
197	0.1	0.16
1110	0.23	0.1
1123	0.17	0.12
1120	0.1	0.17
1117	0.14	0.02
1121	0.13	0.17
1116	0.24	0.17
1115	0.08	0.08
1114	0.08	0.13
1130	0.08	0.13
1132	0.29	0.07
1131	0.11	0.15
1143	0.07	0.09
198	0.8	0.19

Table 3.22 and 3.23 shows the chlorine booster at reservoir and the effect of one booster station on all critical nodes respectively.

Amount of chlorine mass rate to be injected in WDN from every booster station and reservoir is calculated. The formula to calculate the chlorine mass rate is given below.

$M_I = C_I * Q$

M_I = mass rate at injection node corresponding to injection nodes, mg/min.

C_I = chlorine concentration at injection node BS1 to BS11 corresponding to critical nodes, respectively mg/l.

Q = base demand at that node, l/min.

One Sample Calculation Shown Stepwise:

1. Values of Cj are summarized in Table 3.22

2. MI = CI*Q .... Chlorine mass rate to be injected at each node(mg/min).

C_I = is the value of chlorine to be injected in a network and was derived in EPANET.

Q = base demanded at each node and was derived from the network.

Table 3.23: Effect of one booster station on all critical node's summary.

CN\BS	198	1139	1125	1123	1110	194	1115	1112	127	178	1135	RSVR
1105	0	0.26	0	0	0	0.13	0	0	0	0	0	0.16
197	0	0	0	0	0	0.31	0	0	0.01	0	0	0.16
1110	0	0.19	0	0	0.5	0.13	0	0.02	0	0	0	0.1
1123	0	0.08	0.18	0.2	0.21	0.05	0	0.01	0.01	0	0	0.12
1120	0	0	0	0	0	0.16	0	0.13	0.01	0	0	0.17
1117	0	0	0.25	0	0	0.03	0	0.02	0.01	0	0	0.02
1121	0	0	0.33	0	0	0	0	0	0.01	0	0	0.17
1116	0	0	0.29	0	0	0.02	0.1	0.02	0.01	0	0	0.17
1115	0	0	0.08	0	0	0.11	0.5	0.1	0.01	0	0	0.08
1114	0	0	0.08	0	0	0.12	0	0.1	0.01	0	0	0.13
1130	0	0	0	0	0	0.02	0	0.27	0.01	0	0	0.13
1132	0	0	0.08	0	0	0.01	0.3	0.2	0.01	0	0	0.07
1131	0	0	0	0	0	0.01	0	0.3	0.01	0	0	0.15
1143	0	0	0	0	0	0	0	0	0.29	0	0	0.09
198	0.5	0	0	0	0	0	0	0	0	0	0	0.19

3.14 JUSTIFICATION TECHNIQUES

The terrain parameters, i.e., latitude, longitude, and latitude of the study area, are vital for the planning and placement of 8 routes of WTN from Hetwane dam (source) to Dapoli (destination) as well as placement of tank, pipes, and junctions for the WDN at Dapoli. These parameters were acquired through DEM generation in QGIS, and the various possible routes for WTN were considered with the NDVI and EVI analysis. Hence, two justification surveys were conducted to verify these outputs, primarily recording the data through an antenna mounted over a car and secondary by designing a mobile application through ODK.

3.14.1 Mobile Application (ODK)

The results were generated through DEM; a mobile application is made to justify the output through the Open Data Kit (ODK). A field survey has been done using a mobile application, and methodology for data collection is shown in Figure 3.74. This data can be exported in various forms such as tabular form, kml file, graphical representation. By using a mobile application, we had located the existing route of the water transmission line. From the ODK application, we had located coordinates and elevation of water transmission lines at different locations. The survey through ODK collected information at each point of the existing WTN from the existing alignment of the water transmission line (hetawane dam – Tarankhop – kopar- Jite). Table 3.24 shows the data collection entities through the mobile application

Table 3.24: Parameters included in the mobile application for data collection

Parameter	Description	Data type
Reporting person	Name of the reporting Person	Text
Date/time	Date and time of reporting	System-generated
Location	The exact location on the map	GPS coordinates
Image	Image of site	Media file
Video	Video of site	Media file
Pressure	The pressure of each point	Number
Geotagging	Latitude, longitude, altitude	GPS coordinates
Other information	Any relevant information	Text

The entire route was divided into three different sections of approximately six km which are Hetawane to Borgaon, Borgaon to Irwadi, and Irwadi to Jite. Then, filling the blank form in the ODK application was done, adding GPS coordinates of the respective location and supporting the data with photos and videos wherever required. This procedure has been repeated to a different location, and we collected 263 points. Later on, all these data are sent to the cloud server and a water transmission line in google earth by exporting the file from the server in the kml file.

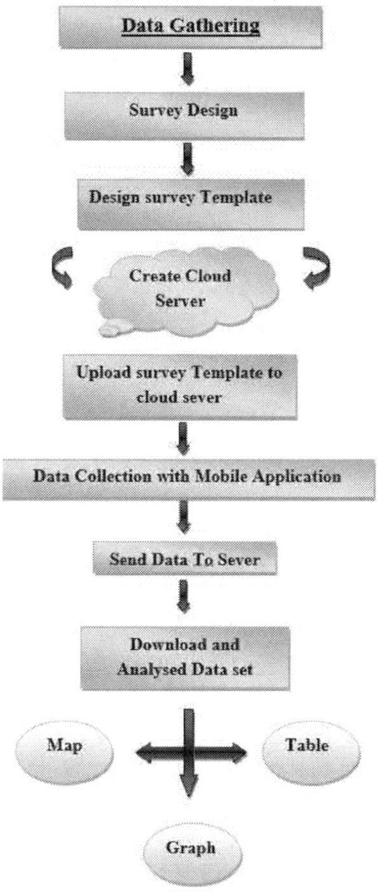

Figure 3.74: Methodology for mobile application through ODK

3.14.2 System Server

The place where accurate data can be stored in one place. The server where all mobiles are connected at one place. The data can be easily manipulated and exported in different formats such as XLS, CSV, ZIP, KML, SAV ZIP, CSV ZIP, Google docs, and XLS Reports. Refer Figure 3.75.

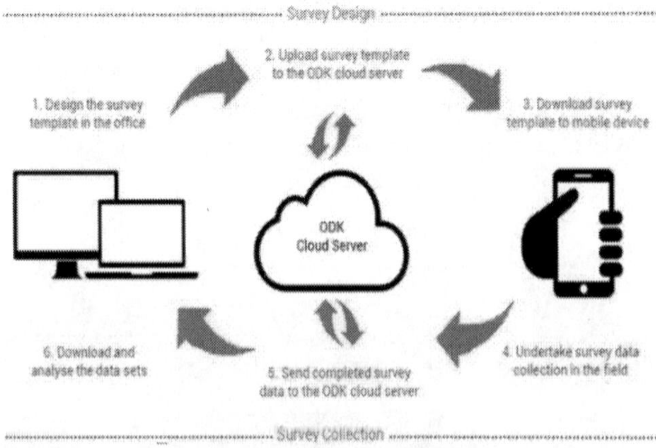

Figure 3.75: System Server
(**Source:** https://geogear.wordpress.com/2016/12/20/qgisodk-new-plugin)

3.14.3 Steps for developing a system server:

- Step 1: Go to http://kf.kobotoolbox.org.
- Step 2: Create a login ID in the kobo toolbox by setting the username and password.
- Step 3: Create a new project. Build a new form, add locations and pictures.
- Step 4: Click on deploy and upload a built form in the ODK. Ref Figure 3.75a

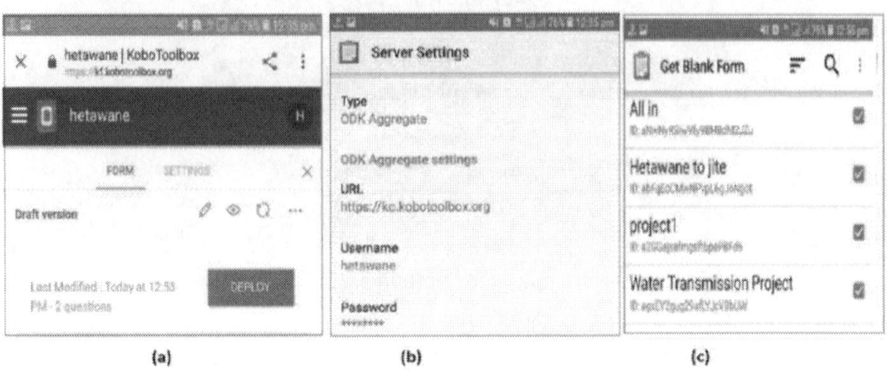

Figure 3.75abc ODK application dropbox
(**Source:** https://geogear.wordpress.com/2016/12/20/qgisodk-new-plugin)

- Step 5: open ODK, collect application go to general setting, server, set URL, Username & Password. Refer to Figure 3.75b

- Step 6: Select getting blank form and download the deployed form we had marked in the server. Refer Figure 3.75c

The survey was done for about 18.8 km. route of the Existing alignment of the water transmission line (hetawane dam – Tarankhop – kopar- Jite), as shown in Figure 3.76

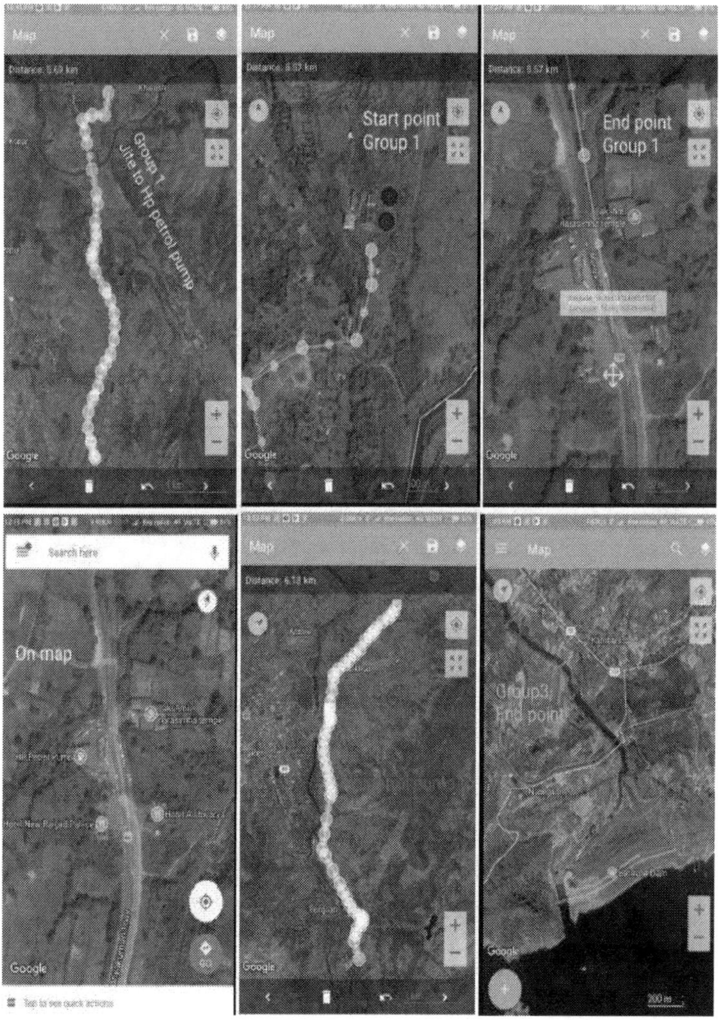

Figure 3.76 Mobile Survey Route

- Step 1: Open the ODK application and select a blank form. Refer to Figure 3.77 a
- Step 2: Add a location of the respective point. Refer to Figure 3.77 b

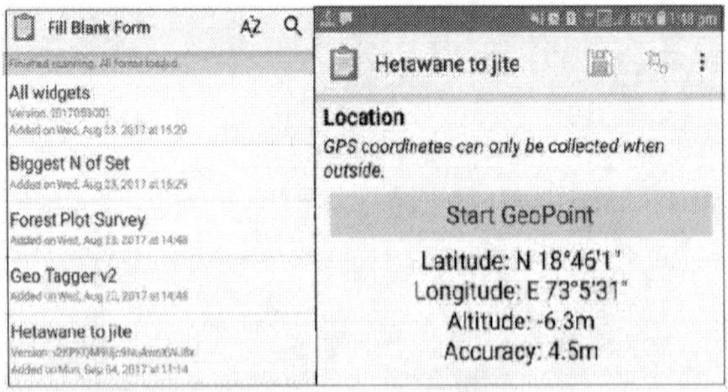

(a) (b)

Figure 3.77: ODK form and application layout

- Step 3: Add a photo with details of the location. Refer to Figure 3.78 a
- Step 4: Describe the pipe. Refer to Figure 3.78 b

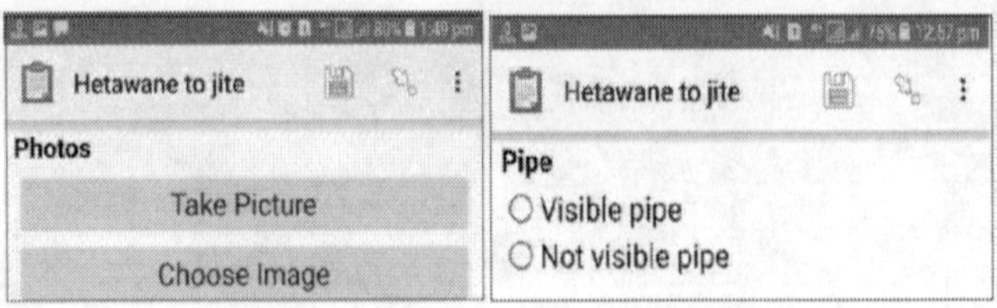

(a) (b)

Figure 3.78: Adding photo and details of pipes in ODK

- Step 5: Name the form and save the filled form. Refer to Figure 3.79
- Step 6: Select getting blank form and download the deployed form, which we have marked in the server. Refer to Figure 3.79

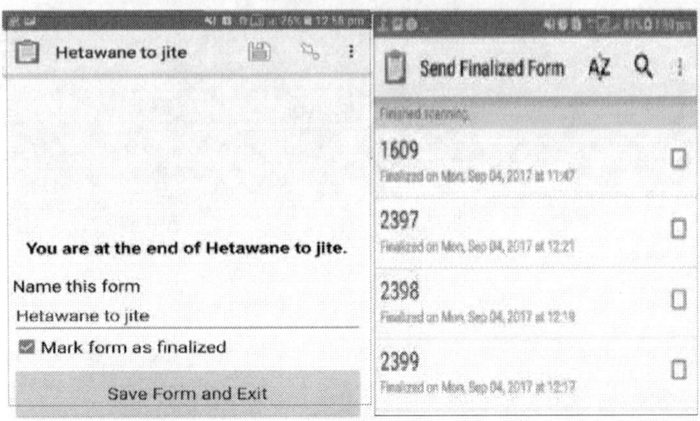

Figure 3.79: Saving and deployment of form

3.14.4 Attributes Elements for Mobile Application:

Different parameters that would be included in the application are the basic attributes, as shown in Table 3.25.

Table 3.25: Attributes elements for mobile application

Parameter	Description	Data type
Reporting person	Name of the reporting Person	Text
Date/time	Date and time of reporting	System-generated
Location	The exact location on the map	GPS coordinates
Image	Image of site	Media file
Video	Video of site	Media file
Pressure	The pressure at each point	Number
Geotagging	Latitude, longitude, altitude	GPS coordinates
Other information	Any relevant information	Text

View of Data Collect application form on mobile is shown in Figure 3.80.

Figure 3.80: View of Data Collect application form on mobile

3.15 GLOBAL POSITING SYSTEM GPS NAVIGATION (GPS)

As the majority of the WTN pipeline runs parallel to NH 66 (Edapally Panvel) and SH 88 (Pen- Khopoli) to the road, an antenna was mounted above the car, and it recorded the latitude, longitude, and altitude of the terrain which lies in the study area as shown in Figure 3.81.

Figure 3.81: Data Collection by an antenna mounted over the car.

This data collected through the antenna helps justify the terrain geography in terms of all the points of WTN (x, y, and z), which later was cross verified with the Digital Elevation Model DEM, which was simulated in the GIS platform.

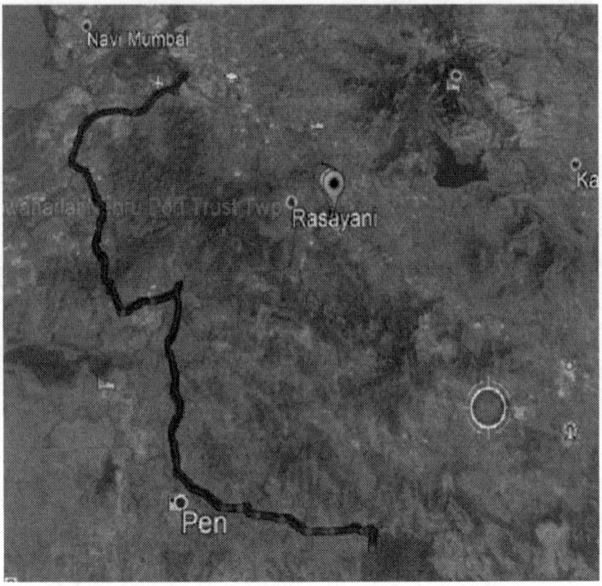

Figure 3.82: Data Collection route by car.

As shown in Figure 3.82, the car traveled over 18.8 km from Hetwane dam until the destination. The antenna was placed over the car and the data was collected using a GPS tracker overhead. As GPS has many advantages compared to traditional methods such as it is weather independent irrespective of fog, rain, snow. have zero effect on the receiving signal. It also does not require line of sight as compared to regular survey requirements. The geodetic accuracy is also high and can be used at night, which is a significant advantage in night surveys and comparatively requires lesser human resources.

Chapter - 04

Results and Discussions

4.1 LAND USE LAND COVER (LULC)

As discussed in previous chapter, various calculations and simulations are done over the objectives of this research, and it is seen that clear and distinct results have arrived towards the end, which this chapter would discuss in detail. With the Landsat image dataset, the NDVI dynamics are understood, and its impact can be analyzed in the Land Use Land Cover (LULC). Figure 4.1 gives a clear picture of keeping the objective of this research intact.

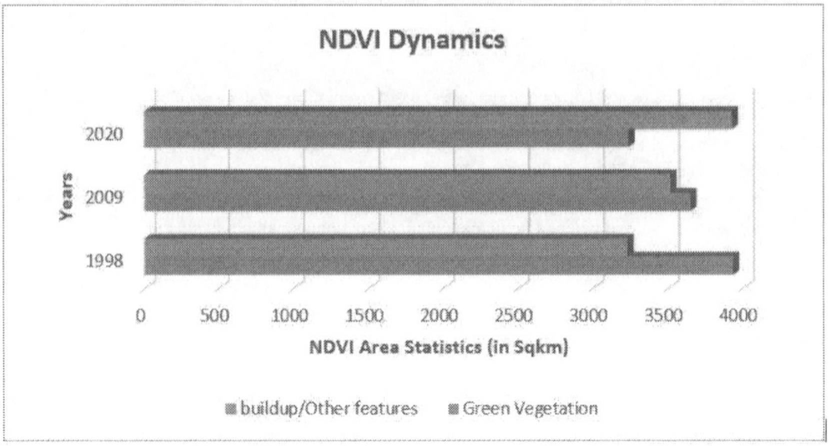

Figure 4.1: Normalized Difference Vegetative Index (NDVI) Dynamics

In Figure 4.2, the agricultural land is converted into a sports city. The road is visible in both the data. (2009 and 2019) with the Landsat image dataset, the NDVI dynamics are understood, and its impact can be analyzed in the Land Use Land Cover (LULC).

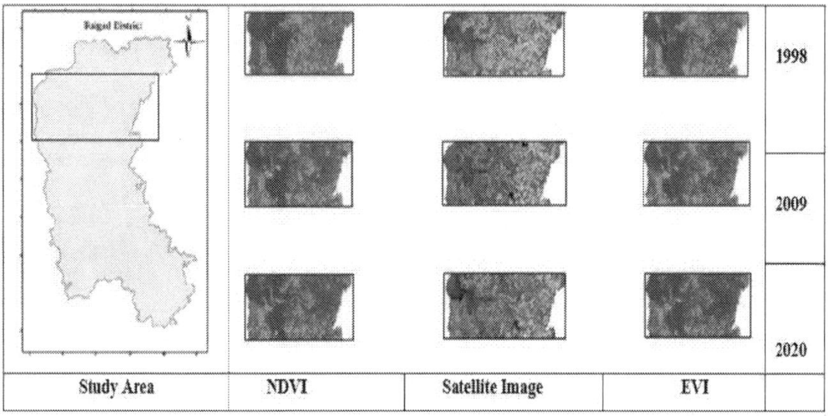

Figure 4.2: Time Series Analysis by Change detection of the study area (1998, 2009, 2020) - 11 years interval

Figure 4.3 gives a clear picture of keeping the objective of this research intact. In Figure 4.4, it can be seen that the LULC has drastically changed, and the urban sprawl is increased in 2020 as compared to 1998. With the Landsat image dataset, the EVI dynamics are understood, and its impact can be analyzed in the Land Use Land Cover (LULC). Figure 4.4 gives a clear picture of keeping the objective of this research intact.

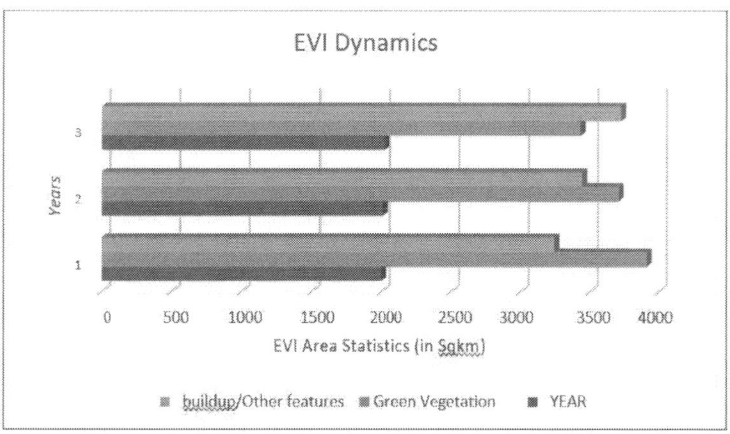

Figure 4.3: Enhanced Vegetative Index (EVI) Dynamics

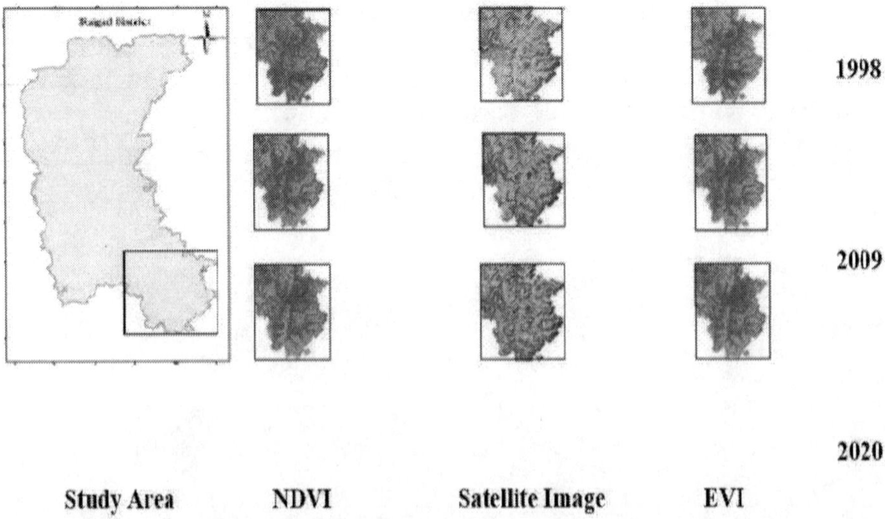

Figure 4.4: Time Series Analysis by Change detection of the study area (1998, 2009, 2020) - 11 years interval

4.2 PARAMETRIC INVESTIGATIONS

The various parametric investigations for WTN, WDN, and Water Quality were conducted and are explained in brief below:

4.2.1 Eight Cases about Water Transmission Network Layout

After a successful run of the model, we can view the analysis results and give all the results such as flow, velocity, unit headloss, pressure. We also get all the results in the graphical format showing flow condition, the pressure at any point in the network, and after analysis of the results of the existing alignment, we can go further for design and analysis of new alignments to find out the best alignment of all the alignment. Refer Figure 4.5. By analyzing case no - 8, we get to know that Gravity flow of 150 LD from the dam to WTP directly is possible with a residual head of 7.92 m at WTP, and Gravity flow of 200 MLD from the dam to WTP is possible with a residual head of 3.54 m at WTP and with a minimum pressure of 1.59 m at node J-66. Table 4.1 shows the comparison of the results for the eight different cases.

By going though all eight cases, alignment for case no. 8 gives valid results and is suggested be the best alternative alignment. In this case, 200 MLD water is reaching a water treatment plant situated at village Jite with a minimum residual pressure of 1.59 meters at node J-66. Therefore, the best route (8th alternative case) is shown in Figure 4.6.

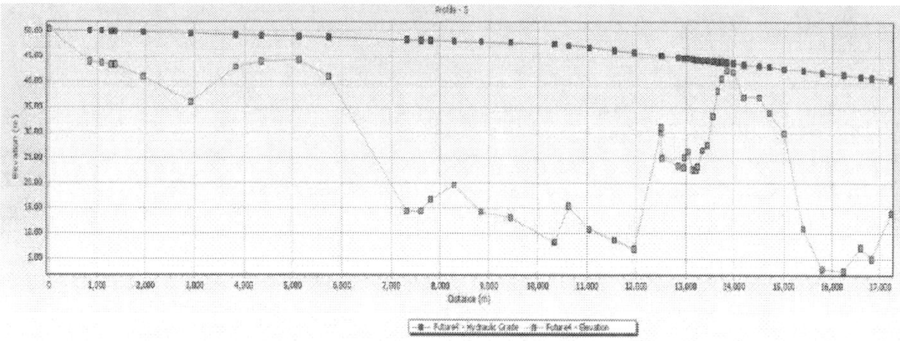

Figure 4.5: L section for case 8

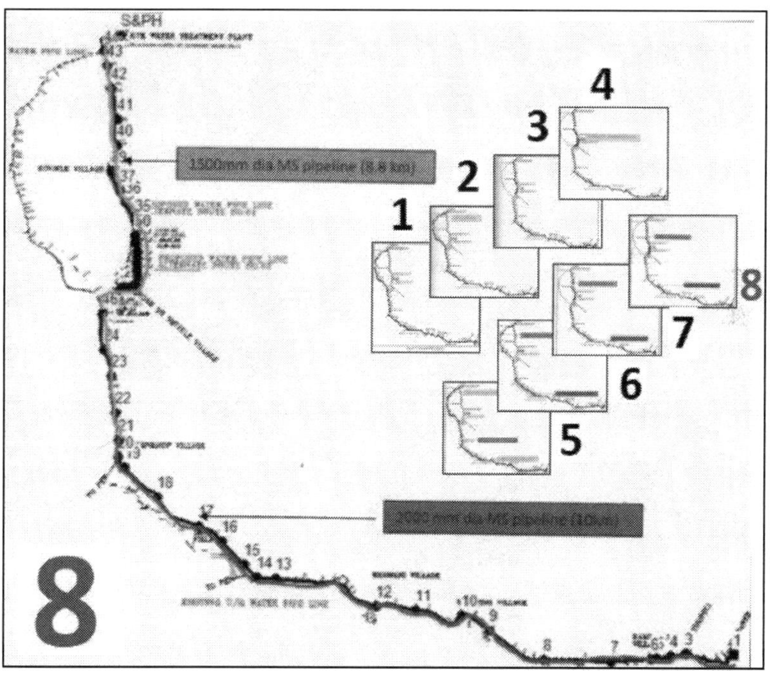

Figure 4.6: Eight alternative cases of Water Transmission Network

Table 4.1: Comparison of the results for eight different cases.

No	Option	Detailed Description	Conclusions	
			Assessment of flow at Sump and pump house with a level of 14m.	**Assessment of flow at WTP, Jite with a level of 36m**
01	Existing alignment (Hetawane Dam – Tarankhope – Kopar – Jite.	A 1500 mm dia PSC line of length 11.90 Km. has been laid from the dam to Eirwadi (Tarankhop), and again a PSC pipeline of 1200 mm dia from Tarankhope to Jite (6.90 km) has been laid. The total length is 18.885 km	It is observed that 150 MLD water reaches the sump and pump house at Jite. However, residual pressure is just 2.57 m at node J-10.	Gravity flow of 150 MLD from the dam to WTP directly is not possible. Water from the sump and pump house would have to be pumped. Hence pumping cost is involved.
02	**Existing alignment** (Hetawane Dam – Tarankhope – Kopar – Jite.	A 1500 mm dia PSC pipeline of 10.00 km has been laid up to Bhogeshwari River and M.S. of 1500 mm diameter (8.80 km) from Bhogeshwari river to WTP been laid. The total length is 18.80 km	(1) It is observed that 150 MLD water reaches the sump and pump house at Jite. However, residual pressure is just 2.57 m at node J-10. (2) It is observed that 200 MLD water reaches the sump and pump house at Jite. However, residual pressure is just 0.31 m at node J-10.	Gravity flow of 150 MLD from the dam to WTP directly is possible with a residual head of 1.73m at WTP. Hence pumping cost is involved.
03	Proposed **Alternative** Alignment I (along NH-17 via Shitole to WTP, Jite)-	In this alignment, 1500 mm diameter- a total of 17.80 km of (M.S. pipeline 6.96 km & PSC 10.416 km) has been considered	(1) For 150 4MLD: Nodes J-40, ... J-46 have negative pressure (Figure 3.6) for about a stretch of 471 m. If this alignment is retained, then tedious tunnel methodology (TBM) shall be adopted with a depth of a minimum of 15m.	Gravity flow of 150 MLD from the dam to WTP directly is not possible. Hence pumping cost
			(2) For 100 MLD: Negative pressures are observed at nodes j-40, ..., J-45 with a minimum of -9.29 m. Hence tedious tunnel methodology (TBM) shall be adopted with a depth of a minimum of 10 m.	

No	Option	Detailed Description	Conclusions	
			Assessment of flow at Sump and pump house with a level of 14m.	Assessment of flow at WTP, Jite with a level of 36m
04	**Proposed Alternative Alignment II** Shitole Ambivali via NH-17 to WTP, Jite-	In this alignment, 1500 mm PSC (10 km and M.S. (8.0 km) total 18.00 km length has been considered	For 150 MLD: It is observed that the negative pressures of -1.18 m prevail at node J-46 and -1.4m at node J-66. Nodes J-46 and J-66 are adjoining nodes. Hence pipes can be laid below 2 m. It is proposed to lay a pipeline between these nodes at 2 m depth so that negative pressure would not occur at these nodes.	Gravity flow of 150 MLD from the dam to WTP directly is not possible. Hence pumping cost is involved.
05	**Future Alignment** Alternative 1:	Considering the 2000 mm dia M.S. pipeline (10 km) & 1500 mm dia MS (1.90 km), and 1200 mm dia PSC (6.90 km)	It is observed that 150 MLD water reaches the sump and pump house at Jite.	Gravity flow of 150 MLD from the dam to WTP directly is possible with a residual head of 0.38 m at WTP. Hence pumping cost is involved.
06	Future Alignment Alternative 2:	Considering the 2000 mm dia M.S. pipe (10.00 km) 1500 mm dia M.S. pipe (8.80 km), a total of 18.80 km	(1) For 150 MLD: It is observed that 150 MLD water reaches the sump and pump house at Jite with min residual pressure of 5.3 m at node J-10. (2) For 350 MLD: It is observed that 350 MLD water reaches the sump and pump house at Jite with min residual pressure of 1.35 m J- 34.	(1) Gravity flow of 150 MLD from the dam to WTP directly is possible with a residual head of 1.73 m at WTP. Hence pumping cost is involved. (2) Gravity flow of 200 MLD from the dam to WTP directly is possible with a residual head of 2.57 m at WTP. Hpumping cost is involved.
07	Future Alignment Alternative 3:	Considering the 2000 mm dia M.S. pipe (10.00 km) 1500 mm dia M.S. pipe (8.80 km), a total of 18.80 km	Nodes J-40, ... J-46 have negative pressure (Figure 3.6) for about a stretch of 471 m. If this alignment is retained, then tedious tunnel methodology (TBM) shall be adopted with a depth of min of 10m.	Gravity flow of 150 MLD from the dam to WTP directly is not possible. Hence pumping cost is involved.

No	Option	Detailed Description	Conclusions	
			Assessment of flow at Sump and pump house with a level of 14m.	Assessment of flow at WTP, Jite with a level of 36m
08	Future Alignment Alternative 4	Considering the 2000 mm dia M.S. pipe (10.00 km) 1500 mm dia M.S. pipe (8.80 km), a total of 18.80 km.	(1) For 150 MLD: It is observed that 150 MLD water reaches the sump and pump house at Jite with a minimum residual pressure of 4.19 m at node J-66. (2) For 200 MLD: It is observed that 200 MLD water reaches the sump and pump house at Jite with a minimum residual pressure of 1.59 m at node J-66.	(1) Gravity flow of 150 MLD from the dam to WTP directly is possible with a residual head of 7.92 m at WTP. (2) Gravity flow of 200 MLD from the dam to WTP directly is possible with a residual head of 3.54m at WTP and a minimum pressure of 1.59 m at node J-66.

4.2.2 Findings from Water Distribution Network

To summarise the principle findings, it can be said that a sustainable WDN with a well-defined layout can be placed considering the elevation of every node and links through DEM with the application of GIS. Furthermore, as mentioned in the methodology, the hypothesis related to the results obtained through hydraulic simulations is well under control as no negative pressure is generated in the network; hence a gravity flow is recommended without using pumps.The summary of the evidence supporting the conclusion is described in graphical and tabular formats in the preceding sections, which state the parameters such as velocity, head, and demand are within their limits and as per the requirements needed for gravity-driven WDN.

The significance of the results has its direct implications as well as can be adopted for practical applications of laying the WDN where the geography of the terrain can be adequately identified through remote sensing and GIS via satellite data, and an optimized network can be placed at a developed site which is under rehabilitation.

The system was run with different pipe materials with different roughness factors, and the results were used to determine the superior pipe material for the water distribution network.It was found that the roughness factor is a deciding criterion in the selection of pipe materials.

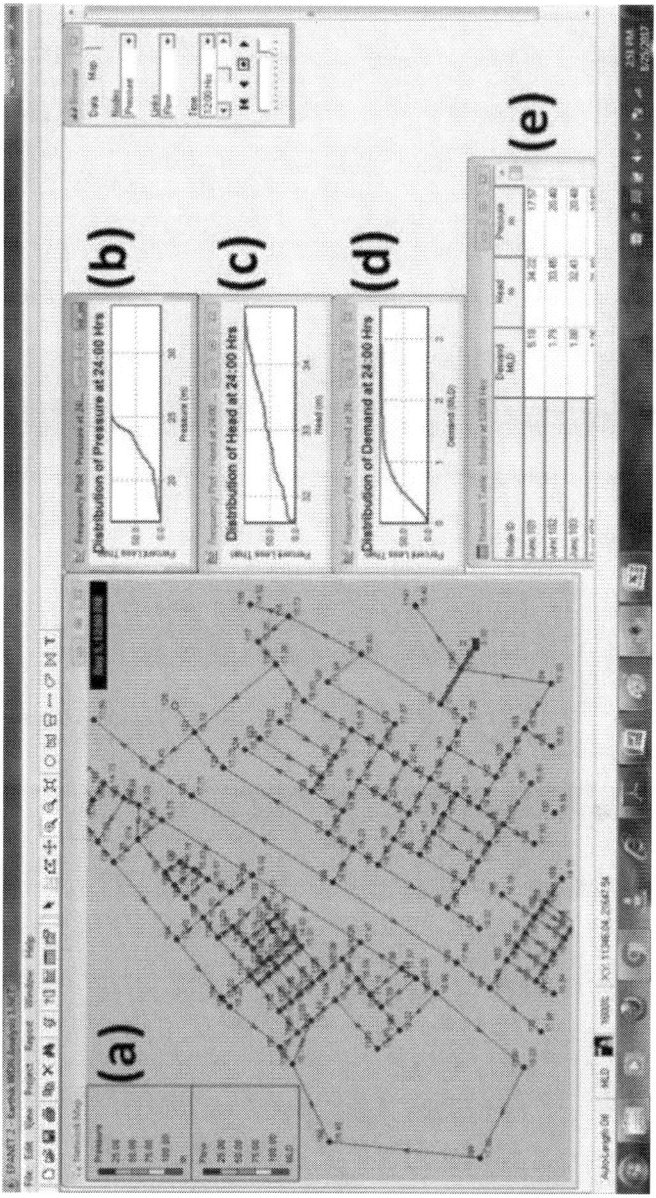

Figure 4.7: (a) Network simulation showing pressure and flow distribution (b) Frequency plots for pressure distribution (c) Frequency plots for head distribution (d) Frequency plots for demand distribution (e) Nodal Table showing demand head, and pressure.

Furthermore, this study showed that higher optimization could be achieved by analysis and repetitive methods by altering pipe diameters. Figure 4.7 shows (a)network simulation showing pressure and flow distribution (b) Frequency plots for pressure distribution (c) Frequency plots for head distribution (d) Frequency plots for demand distribution (e) Nodal Table showing demand head, and pressure.This research can be extended ahead in obtaining other parameters such as leakages, non-renewable water NRW if high-resolution satellite data (other than LISS III & CARTOSAT I) is compared with the current scenario by providing a sTable platform for analyzing the causes and effects on operation and maintenance of WDN.

4.2.3 Water Quality

According to the results, the total amount of chlorine mass rate to be injected in WDN is 26850.93 mg/min from all booster stations and reservoirs to gratify the limitation of 0.2 mg/l.

As per IS 10500 (2012), the initial quality of water is set in the software, and it is run for 2 hrs. Figure 4.8 shows the booster station position in WDN. So, from the simulation in EPANET, we have come to know which nodes can be used to place booster stations as injection nodes. Optimization is done further to know what is the effect of each

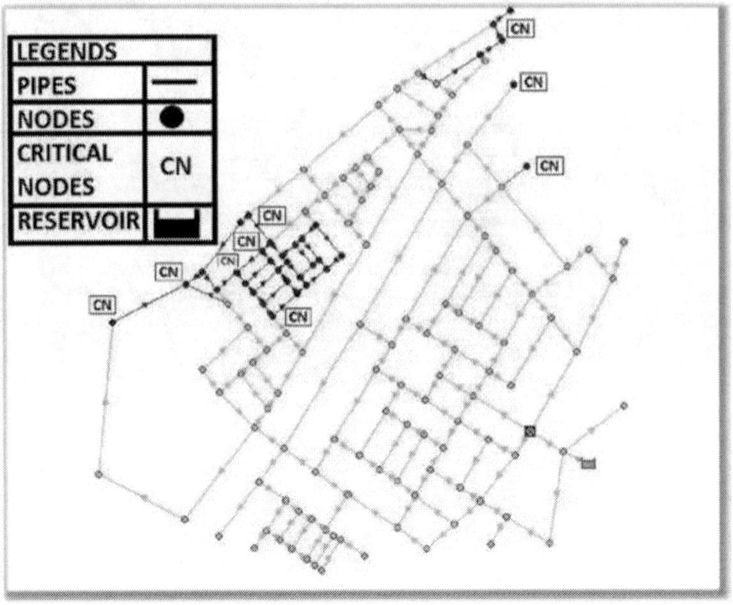

Figure 4.8: Booster Station (BS) position in WDN

booster station on other critical nodes, i.e., (when only one booster station is active as chlorine injection source and all other booster stations are closed along with reservoir) and to know the amount of chlorine mass rate is to injected in WDN. Calculation of chlorine mass rate to be injected is shown below and in detail in Table 4.2.

M_I(mg/min) for node 198 is calculated as follows

The value of MI(mg/min) is calculated by multiplying C_I (mg/l) by Q(l/min): Q(l/min) =Q(MLD)*(106/24*60).

555.2(l/min) =0.8(MLD)* (106/24*60).

M_I(mg/min) =0.5(mg/l) * 555.2(l/min).

=277.76(mg/min).

Table 4.2: Chlorine mass rate to be injected.

Injection node or Booster station.	CI (mg/l).	Q = Base demand (l/min).	$M_I = C_I*Q$ Chlorine mass rate to be injected at each node (mg/min).
M1 Node 198	0.5	555.2	277.76
M2 Node 1139	0.5	624.96	312.48
M3 Node 1125	0.4	236.96	94.4384
M4 Node 1123	0.2	118.048	23.6096
M5 Node 1110	0.5	159.712	79.856
M6 Node 194	0.4	0	0
M7 Node 1115	0.5	55.562	27.776
M8 Node 1112	0.4	236.872	94.7488
M9 Node 127	0.4	361.088	144.4352
M10 Node 178	0.4	159.712	63.8848
M11 Node 1135	0.4	41.664	16.6656
MO Reservoir	0.5	51430.556	25715.28
TOTAL			26850.9344

4.4 JUSTIFICATION TECHNIQUES

As discussed in the earlier chapter, the mobile application which has been created has validated as below:

4.4.1 Mobile Application (ODK)

The whole long route of the water transmission network, which is about 18.8 km long, was geotagged on the google earth platform using the free, open-source tool ODK and near about 261 locations were traced out as it can be seen in Figure 4.9.

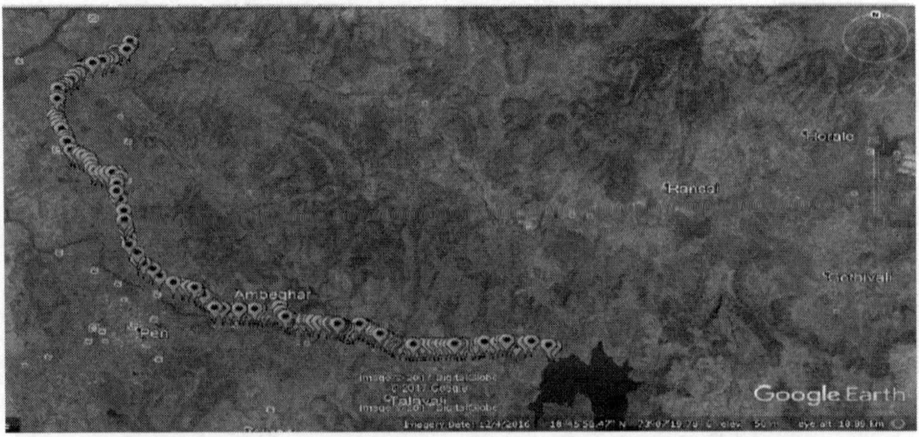

Figure 4.9: Alignment of pipeline in google earth.

Once the Mobile application is made, various details such as reports, Tables, galleries, downloads, and maps are made available for this research work. The results of the said route are filled in a designed template form, and the fundamental data is available in various downloadable formats. Therefore, the time required was less as compared to the traditional survey.

As shown in Figure 4.10, various attributes are recorded while tracking the WTN throughout the source to the destination. The data was recorded with exact location in terms of latitude, longitude, altitude, and precision of about 3 to 5m. All the photos and videos taken have a unique number by which the entity's identification becomes easy. A separate entry is made in the mobile application form wherein we can fill in the pipe which we are captured is visible or not, as shown in Figure 4.11

It was seen that about 192 pipes were not visible, whereas 71 were visible. The visibility reduces due to underground pipe networks and many time due to alignment of pipelines through fields and forest. Figure 4.43 and 4.44 shows visible and non-visible pipelines along the study area respectively

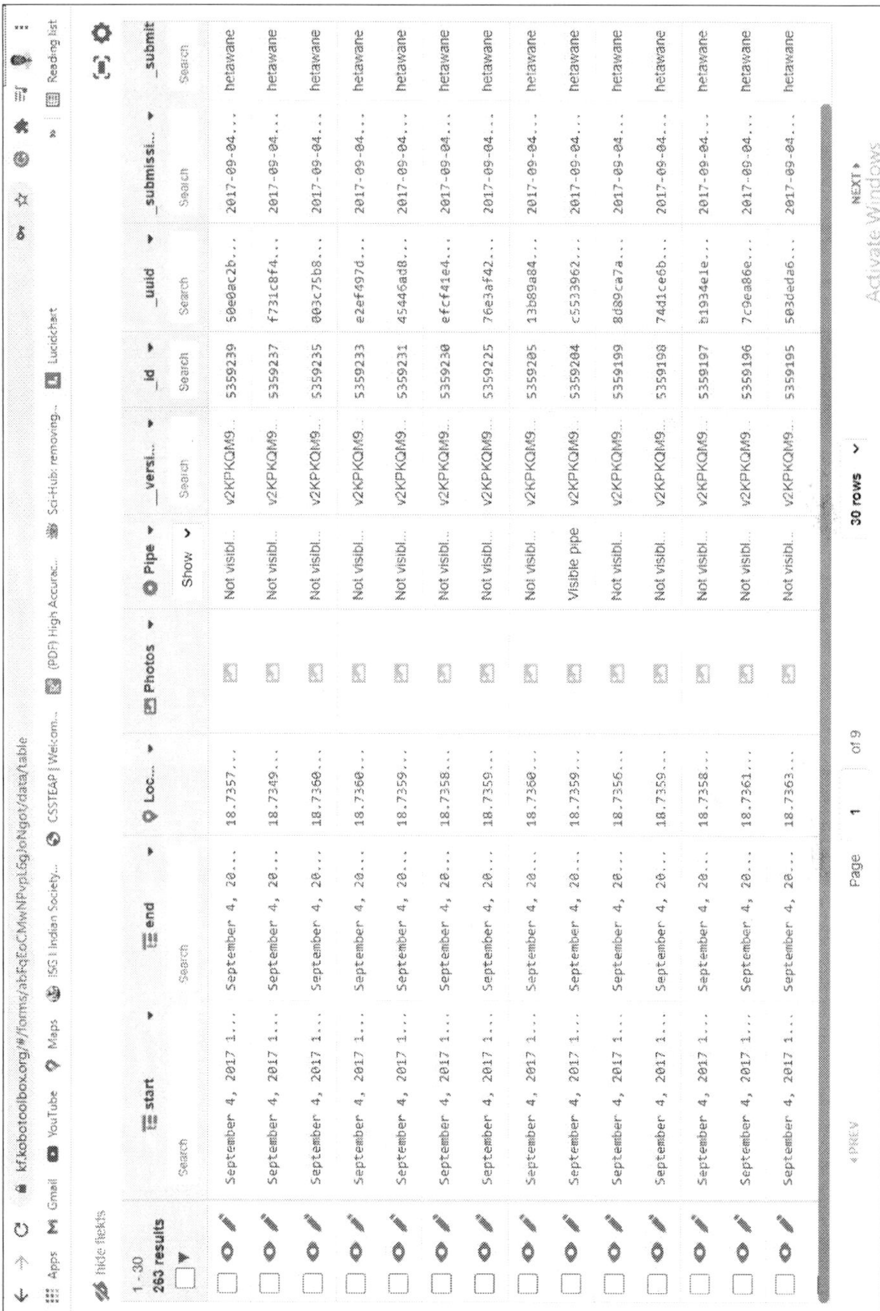

Figure 4.10: Mobile Data in Tabular Formats

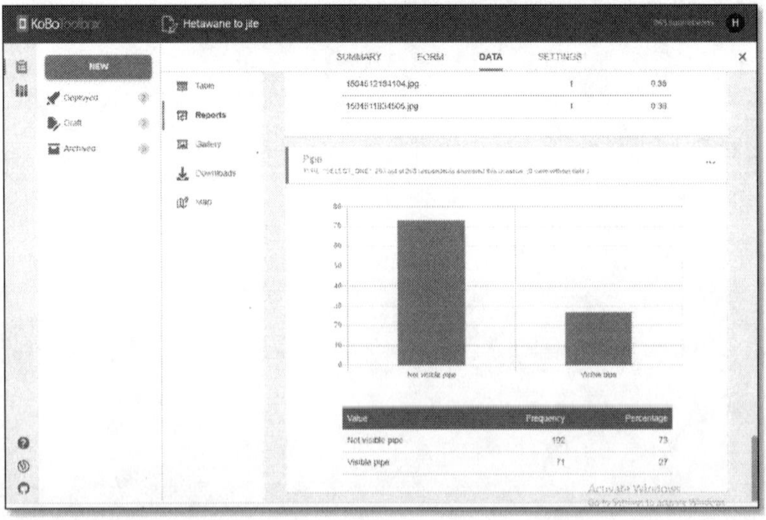

Figure 4.11: Pipeline invisible or not visible (Graphical Output)

CHAPTER - 05

Conclusion and Future Scope

5.1 CONCLUSION

This study emphasizes and helps to understand the ground elevation profile, which can be further used to lay the Water Transmission Network (WTN) and Water Distribution Network (WDN), respectively, keeping in mind the ground slope to water the flow gravity.

In NDVI and EVI research area statistics and image, interpretation is developed both in statistical and visual form for ease, and the results state a decrease in Normalized Difference Vegetation Index of 45.2 % (2020) and 50.9 % (2009) compared to 54.9 % (1998) while increase in built-up features index of about 54.8 % (2020) and 49.1 % (2009) % compared to 45.1 % (1998) respectively. It can also be concluded that decerased in Enhanced Vegetation Index to 47.9 % (2020) and 51.9 % (2009) compared to 54.7 % (1998) while increase in built- up features index of about 52.1 % (2020) and 48.1 % (2009) compared to 45.3 % (1998) respectively.

In the water transmission network, by analysing eight different alternative cases, it is seen that in case no - 8, it is seen that the gravity flow of 150 LD from the dam to WTP directly is possible with a residual head of 7.92 m at WTP and Gravity flow of 200 MLD from the dam to WTP directly is possible with a residual head of 3.54 m at WTP and with a minimum pressure of 1.59 m at node J-66. Therefore, it is recommended that case 8 may be implemented. To summarize the findings of the water distribution network, it can be said that a sustainable WDN with a well-defined layout can be placed considering the elevation of every node and links through DEM with the application of GIS. All hydraulic parameters such as head, pressure, discharge, pipe diameter calculated are satisfied and within specified limits. As mentioned in the methodology, the hypothesis related to the results obtained through hydraulic simulations is well under control as no negative pressure is generated in the network; hence a gravity flow is recommended without using pumps.

In water quality simulation, we have learned which nodes can place booster stations as injection nodes. As a result, the total amount of chlorine mass rate to be injected in WDN is 38665.3455 gm/day from all booster stations and reservoirs to satisfy the constraint of 0.2 mg/l.

A mobile application is made using the free, open-source tool Open Data Kit (ODK), and about 261 locations were traced. Moreover, the whole long route of the water transmission network, which is about 18.8 km long, is geo-tagged on the google earth platform for justification.The recorded data which was collected is later cross-verified through the Global Positioning System GPS antenna over a car which helped in justifying the terrain geography in terms of all the points of WTN (x, y, and z), which later was cross verified with the Digital Elevation Model DEM which was simulated in a GIS platform.

5.2 FUTURE SCOPE

Finally, it can be concluded that there is the immense future scope in this research, especially by using high-resolution satellite images and simulating them along with water supply networks and other infrastructural facilities of the cities. Booster chlorination cab is applied through sensors at various levels across networks. High-level satellite simulated maps can be generated to prepare many well-defined Digital Elevation Models. Unmanned Arial Vehicles (UAVs) Drones can be used for collecting geographical data with more accuracy. Through coding and python language, a touch screen mobile application can be made wherein through zoom in and out the images on the mobile screen, the attributes of every node can be identified by which the administration can work fast towards achieving the operation and maintenance. Few more areas where more research can be carried out are as follows:

1. Using multispectrical and hyperspectral satellite data.

2. Creating user friendly mobile applicationswhich can be conneted to main server by high speed internet.
3. Amount and method of application of chlorine to water distribution network utilizing secondary disinfection.
4. An entirely liable method to identify free bacteria concentration in water distribution systems.

Bibliography

1. A Saminu, Abubakar, Nasiru, L Sagir(2013) “Design of N.D.A. Water Distribution Network Using E.P.A.N.E.T.” International Journal of Emerging Science and Engineering (I.J.E.S.E.) ISSN, 1(9), July 2013, pp 2319–6378
2. A Vasan and Slobodan P. Simonovic, “Optimization of Water distribution network design Using Differential Evolution” Journal of Water Resources Planning and Management © Asce / March/April 2010 / 279-287.
3. A.Antonowicz, R.Brodziak, J.Bylka, J.Mazurkiewicz, S.Wojtecki1, P.Zakrzewski.(2018), “Use of EPANET solver to manage water distribution in Smart City” Water, Wastewater and Energy in Smart Cities, 1-8.
4. Abhishek Lowe, Karthik Nagarajan and Raju Narwade (2020): “Estimation of Rate and Time Delay for underground Mumbai Metro Corridor by using Monte Carlo Simulation “, International Journal of Management and Humanities (IJMH), (June 2020) ISSN Print: 2394 –0913, Volume-4, Issue-10, pp: 68-77.
5. Aditya Shatri, Karthik Nagarajan, RajuNarwade and (2019): “Integrated Land-Use Zoning, Using Topographical Data: Optimizing Vacant Space For Urbanization At Akole Taluka, Maharashtra, India” International Journal of Advanced Research in Engineering and Technology (IJARET), Volume 10, Issue 1, (Jan-Feb 2019), pp. 188-199, IAEME Publication, Article ID: IJARET_10_01_018, Print: 0976-6480 and ISSN Online: 0976-6499
6. Akshata Mestry, Karthik Nagarajan, Raju Narwade(2020): “Estimation of Water Balance components of Watersheds in the Manjira River Basin Using Swat Model and GIS”, International

Journal of Engineering and Advanced Technology (IJEAT), ISSN: 2249- 8958, Volume-9, Issue-3, (February 2020).

7. Akshata Patel, Karthik Nagarajan, Dr. J. E. Nalavade, Archana, Arudkar, Rohan Kumar (2020) "Mobile App for Rural Development and Governance of Village Centric System – MCDM Approach", International Journal for Scientific Research & Development (IJSRD), (April 2020), ISSN (online): 2321-0613, Volume-07, Issue-12, pp-775-777, April 2020.
8. Alperovits, E., Shamir, U., 1977. Design of optimal water distribution systems.Water Resources Research 13 (6), 885-900.
9. Alvisi, S., Franchini, M., Marinelli, A., 2007. A short-term, pattern-based model for water-demand forecasting. Journal of Hydroinformatics 9 (1), 39-50.
10. Amit Kumar Sharma, Karthik Nagarajan and Raju Narwade (2021): "Development of Solar Concentrator studded with Optical Fiber for Daylighting", International Journal of Scientific and Research Publications, (Nov-2021) ISSN: 2250-3153, Volume-11(Issue-11), pp.527-541.
11. Amit Kumar Sharma, Karthik Nagarajan and Raju Narwade(2021): "Development of Solar Concentrator studded with Optical Fiber for Day lighting",International Journal of Scientific and Research Publications, (Nov-2021) ISSN: 2250-3153, Volume-11(Issue-11), pp.527-541.
12. Anand Upadhyay, Santoshkumar Singh, Shailesh Kisan Gaikwad, and Ashutosh Chandra Mukherjee.(2017), "Classification and Comparative Study of IRS LISS-III Satellite Images Using R.B.F.N. and Decision Tree "Springer Nature: Sensors and Image Processing,245-253
13. Arfanuzzaman Md., Atiq Rahman A. 2017. Sustainable water demand management in the face of rapid urbanization and groundwater depletion for social-ecological resilience building. Global Ecology andConservation 10 (2017) 9–22.
14. Armando Di Nardo, Michele Di Natale, Carlo Giudicianni, Roberto Greco and Giovanni Francesco Santonastaso(2017), "Complex network and fractal theory for the assessment of water distribution network resilience to pipe failures," I.W.A. Publishing 2017:water science and technology: water supply ws2017124, DOI: 10.2166/ws.2017.124

15. Arya Vijayan, Karthik Nagarajan, Raju Narwade(2019): "Real-Time Water Leakage Monitoring System Using IoT Based Architecture " International Journal for Research in Engineering Application & Management (IJREAM), (Nov-2019), ISSN: 2454-9150, Vol-05, Issue-08, pp.24-30.

16. Austin, R. G., Choi, C. Y., Balagopal, T. C., and Gurian, P. L. "Event-based microbial risk assessment and response analysis of Cryptosporidium in municipal water distribution networks." Water Distribution Systems Analysis (2010).

17. Baiyi Jiang, Fabei Zhang, Jian Gao, and Hongbin Zhao, (2012)" Building Water Distribution Network Hydraulic Model by Using WaterGEMS" ASCE, International Conference on Pollution and Treatment Technology, I.T.C.P.T.T., Vol1, Oct 2012, pp 453-461

18. Behzadian, K., Kapelan, Z., Savic, D., Ardeshir, A., 2009. Stochastic sampling design using a multiobjective genetic algorithm and adaptive neural networks, Environmental Modelling & Software 24 (4), 530-541.

19. Betanzo, E. W., Hofmann, R., Hu, Z., Baribeau, H., and Alam, Z. "Modeling the Impact of Microbial Intrusion on Secondary Disinfection in a Drinking Water Distribution System." Journal of Environmental Engineering (2008).

20. Biradar Shilpa, Karthik Nagarajan, Raju Narwade, Gayatri Deshpande,(2019): "E-Waste: An Alternative to Partial Replacement of Coarse Aggregate in Concrete", International Journal Of Engineering Research & Technology (IJERT), (July-2019) ISSN: 2278-0181, Volume 08, Issue 07 (July 2019), pp.993-999

21. Brumbelow, K., Torres, J., Guikema, S., Bristow, E., Kanta, L., (2007)." Virtual Cities for Water Distribution and Infrastructure System Research" World Environmental and Water Resources Congress, Restoring Our Natural Habitat, Tampa. Florida.

22. Cascón-Katchadourian, Jesús-Daniel; Ruiz-Rodríguez, Antonio-Ángel (2016). "Descripción y valoración del software MapTiler: del mapa escaneado a la capa interactiva publicada en la Web". El profesional de la información, v. 25, n. 6, pp. 970-978.

23. Chary, 2005, "24 Hour Water Supply: A Goal Achievable", ASCE, Nagari, ASCI, Vo 1, pp 22-23

24. Chhaya Zende, Karthik Nagarajan, RajuNarwade(2019): "Manufacturing Techniques of Sustainable Recyclable Formwork By A Smart Material Waste Composite Material (WCM) for

Infrastructural Projects of Future Cities - A Swachh Bharat Abhiyan Initiative", International Journal of Management, Technology And Engineering (IJMTE), (June-2019), ISSN NO: 2249-7455, Vol:9, Issue:6, pp.666-671

25. Chirag Dhoble, Sanskruti Pashte, Karthik Nagarajan, RajuNarwade, (2020): "Pressure based Air Purification Lamp for Multifunctioning Purpose" International Research Journal of Engineering and Technology (IRJET)", (Aug-2020) Vol:7, Issue:8, e-ISSN: 2395-0056, p-ISSN: 2395-0072, pp.2471-2479
26. Chowdary Mohanlal, Karthik Nagarajan, RajuNarwade(2019): "Applications of 4D GIS Model in Construction Management ", International Journal of Innovative Technology and Exploring Engineering (IJITEE), (July-2019) ISSN: 2278-3075, Volume-8 Issue-9, July 2019, pp.2597-2608
27. Chyr Pyng Liou. (1998)."Limitations and Proper Use of the Hazen-Williams Equation" ASCE, Journal of Hydraulic Engineering 124(9),951-954.
28. Crespo-Sanz, Antonio; Fernández-Wyttenbach, Alberto (2011). "¿Cartografía antigua o cartografía histórica?". Estudios geográficos, v. 72, n. 271, pp. 403-420.
29. Dandy, G.C., Simpson, A.R., Murphy, L.J., 1996. An improved genetic algorithm for pipe network optimization. Water Resources Research 32 (2), 449-458.
30. Darshan Mehta, Krunal Lakhani, Divy Patel, Govind Patel (2015), "Study Of Water Distribution Network Using Epanet," iJARESM, international journal of advanced research in engineering and management, ISSN: 2394-1766.
31. Decision Making in Civil Engineering,11-13 October 1998, Montreal, Canada.
32. DeFries, R., Hansen, M., 1995. Global discrimination of land cover types from metrics derived from A.V.H.R.R. pathfinder data. Remote Sens. Environ. 54 (3), 209 222. https://doi.org/10.1016/0034-4257(95)00142-5.
33. Dhondabai S Narayankar, Karthik Nagarajan and Raju Narwade(2020): "Soil Erosion Modelling using RUSLE and G.I.S. of Dehrang Dam Watershed, Panvel, District Raigad, Maharashtra" G.I.S. Science Journal (G.S.J.), ISSN: 1869-9391, Volume-7, Issue-6, pp-732-743, (June 2020).
34. Dhondabai S Narayankar, Karthik Nagarajan, Raju Narwade (2020): "Soil Erosion Modelling using RUSLE and GIS of Dehrang

Dam Watershed, Panvel, District Raigad, Maharashtra" GIS Science Journal (GSJ), ISSN: 1869- 9391, Volume-7, Issue-6, pp-732-743, (June 2020).

35. Di Pierro, F., Khu, S.T., Savic, D., Berardi, L., 2009. Efficient multiobjective optimal design of water distribution networks on a budget of simulations using hybrid algorithms. Environmental Modelling & Software 24 (2), 202-213.

36. Divyashree Yadav, Karthik Nagarajan, Raju Narwade, Poonam Tiwari, Heena Pande (2020): "Automatic Urban Road Extraction from High-Resolution Satellite Data Using Object-Based Image Analysis: A Fuzzy Classification Approach", Journal of Remote Sensing & GIS, (Dec-2019), ISSN: 2469-4134, Volume-09, Issue-4,pp-1-8, (Aug 2020).

37. Douglas Engoke Musiega, David Ndegwa Kuria, Moses Murimi Ngigi and Simon Kibue Ngugi (2012), "An open-source Geographic Information System (G.I.S.) approach to water supply management, distribution, and billing" International Journal of Computer Engineering Research Vol. 3(4), pp. 55-62, November 2012.

38. Environment South Bank University,2000, pp 1-10, 202, U.K.

39. Evan Newcomer, Courtney Boyd, Laban Nyirenda, Emmanuel Opong, Shannon Marquez and Rochelle Holm "Reducing the burden of rural water supply through greywater reuse: a case study from northern Malawi."

40. Farzaneh Baghal Asgharia, JalilJaafarib, MahmoodYousefia, Ali Akbar Mohammadi, and Reza Dehghanzadehd. (2018), " Evaluation of water corrosion, scaling extent and heterotrophic plate count bacteria in asbestos and polyethylene pipes in drinking water distribution system " Human and Ecological Risk Assessment: An International Journal, 1138-1149.

41. G.I.S. techniques for Budigere Amanikere watershed, Dakshina Pinakini Basin, Karnataka, India" Applied water science 2017, 7, (8),4399–4414

42. Geem, Z. W.& Cho, Y. H., (2011) "Optimal Design of Water Supply Networks Using Parameter-Setting-Free Harmony Search for Two Major Parameters," ASCE, Water Resources Planning and Management,137 (4), July 2011, pp 377-380

43. Giustolisi, O., Savic, D., 2010. Identification of segments and optimal isolation valve system design in water distribution networks. Urban Water Journal 7 (1), 1-15.

44. Goldman Sachs Economic Research, The European Flood Risk Directive [E.U.,2007]

45. Goyal Roopali V. & H.M. Patel "Optimal location and scheduling of booster chlorination stations for drinking water distribution system "Journal of Applied Water Engineering and Research (2016).

46. Grayman, SudhirKshirsagar, Melixa Rivera-Sustache landmark Ginsberg"An Improved Water Distribution SystemChlorine Decay Model Using EPANET MSX." Journal of Water Management Modelling R245-21 (2007).

47. Gunhui Chung (2007), "Water Supply System Management Design and Optimization Under Uncertainty," The University of Arizona.

48. Hart.W., and Murray. R, "Review of Sensor Placement Strategies for Contamination Warning Systems in Drinking Water Distribution Systems." Water Resource Planning Manage (2004).

49. Held, A.A., 2019. Mapping the multi-decadal mangrove dynamics of the Australian coastline. Remote Sens. Environ.

50. Hyndman R, Verbesselt, J.,Newnham, G., Culvenor, D., 2010. Detecting trend and seasonal changes in satellite image time series. Remote Sens. Environ. 146, 106–115. https://doi.org/10.1016/j.rse.2009.08.014.

51. Indian Standard (IS-10500:2012) Drinking Water Specification (Second Revision).

52. Ingeduld P., Pradhan, A., Svitak, Z., Terrai, A. (2008)" Modelling Intermittent Water Supply Systems with E.P.A.N.E.T.," ASCE, Water Dist. Systems Analysis Symposium,1-8.

53. J. Duchon, 1976, Splines minimizing rotation-invariant semi-norms in Sobolev spaces. pp 85–100, In Constructive Theory of Functions of Several Variables, Oberwolfach 1976, W. Schempp and K. Zeller, eds., Lecture Notes in Math., Vol. 571, Springer, Berlin, 1977. doi:10.1007/BFb0086566

54. Jensen, J.R., 2005. Chapter 12, Digital Change Detection. Introductory Digital Image Processing: A Remote Sensing Perspective, third. Pearson Prentice Hall, Upper Saddle River, New Jersey, pp. 467–494 ISBN:0131453610.

55. Ji Soo Lee, Won Hwa Hong, (2014)"Development of small district adjustment on public water demand to decrease municipal water supply load," Urban Water,13(2), Sept.2014, pp 142-155

56. Jiménez-Pelayo, Jesús; Bonachera-Cano, Francisco-José (2001). "Recursos de información cartográfica en internet". En: Jiménez-Pelayo, Jesús; Monteagudo- López-Menchero, Jesús. La documentación cartográfica: tratamiento, gestión y uso. Huelva: Universidad de Huelva, pp. 221-262. ISBN: 84 95699 03 6

57. Jolly, M., Lothes, A., Sebastian Bryson, L., Ormsbee, L.(2014)." Research Database of Water Dist. System Models" ASCE, Water Resources Planning & Mgt,140 (4)

58. Justice, C.O., Townshend, J.R.G., Vermote, E.F., Masuoka, E., Wolfe, R.E., Saleous, N., Roy, D.P., Morisette, J.T., 2002. An overview of M.O.D.I.S. land data processing and product status. Remote Sens. Environ. 83 (1–2), 3–15. https://doi.org/10.1016/ S0034-4257(02)00084-6.

59. K. Vairavamoorthy a,b, Jimin Yana, Harshal M. Galgale a, Sunil D. Gorantiwar, (2006), "R16 IRA-WDS A GIS-based risk analysis tool for W.D. system", Elsevier, Environmental Modelling & Software, Vol 22 7, Jan 2007, pp 951-965

60. Kalanithy Vairavamoorthy" Design of Sustainable Water Distribution System In Developing" Water Development Research Unit, Faculty of the Built

61. Kalungi, P., Tanyimboh, T.T., 2003. Redundancy model for water distribution systems. Reliability Engineering & System Safety 82 (3), 275-286.

62. Karthik Nagarajan (2019) : "Case Study of Skill Development and Capacity Building of Space Science in Rural Areas by Pillai H.O.C. College of Engineering and Technology (PHCET) - A IIRS ISRO Outreach Network Institute," National Symposium on Innovations in Geospatial Technology for Sustainable Development with special emphasis on N.E.R. Shillong, November 20-22, 2019, Meghalaya, India

63. Karthik Nagarajan (2020): "Education, Training and Capacity Building in spreading awareness of Remote Sensing, G.I.S. and GeopSpatial Technology in a Rural Village in Maharashtra through I.I.R. Outreach Program by Pillai H.O.C. College of Engineering and Technology – A Well Analysed Case Study," Asian Conference of Remote Sensing, Deqing, (Nov 2020) 10-12, 2020.

64. Karthik Nagarajan (2020): "Education, Training and Capacity Building in spreading awareness of Remote Sensing, G.I.S.

and GeopSpatial Technology in a Rural Village in Maharashtra through I.I.R. Outreach Program by Pillai H.O.C. College of Engineering and Technology – A Well Analysed Case Study“ Asian Conference of Remote Sensing, Deqing (Nov 2020) 10-12, 2020.

65. Karthik Nagarajan and Sunilkumar Patel (2017):“ Sustainable Smart Blue Roof Network System with an application of Geographic Information System (G.I.S.) “ at International Conference of Emerging Technology in Structural Engineering (ETSE-2017): ISSN (Online) 2456-1290,Vol-2, Issue:3, pp.423-428, March 2017

66. Karthik Nagarajan and S.B.Charhate (2017) “Application of Geographic information system for water distribution networks through Quantum G.I.S. plug-in with hydraulic simulation for infrastructure and development planning” The 38th Asian Conference of Remote Sensing A.C.R.S., (Oct 2017), New Delhi

67. Karthik Nagarajan and S.B.Charhate (2017) “Computational based simulation considering various pipe materials in design and analysis of water distribution networks “ at 22nd International Conference on Hydraulics, Water Resources and Coastal Engineering- HYDRO 2017, Dec 2017, held at L.D. College of Engineering, Ahmedabad from21th and 23rd (Dec 2017)

68. Karthik Nagarajan and S.B.Charhate (2018) “Water Distribution Network: A Remote Sensing and G.I.S. Approach” at Next Frontiers In Civil Engineering: Sustainable and Resilient Infrastructure, held at I.I.T. from November 30 to Dec 2018

69. Karthik Nagarajan and Shrikant Charhate (2016): “Smart Modal Analysis of Multistoried Building Considering the Effect of Infill Walls.”, International Journal of Global Technology Initiatives(IJGTI) 5.1 (March 2016), Vol:5, Issue:1,p-ISSN: 2277-6591,e-ISSN: 2320-1207 pp.C20-C27.

70. Karthik Nagarajan and Shrikant Charhate (2016):“Smart Modal Analysis of Multistoried Building Considering the Effect of Infill Walls “ at “5th International Conference on Global Technology Initiatives”ISSN (Print) 2277-6591 & ISSN (Online) 2320-1207,Vol 5, Issue 1, April 2017, pp.C20-C27.held at Rizvi College of Engineering, Mumbai on 29th and 30th (March 2016)

71. Karthik Nagarajan, Manav Dodiya, Akshay Chavan, Sagar Davate,Sameer Prajapati (2019) : “Sustainable Water Harvesting

Technique by Condensation of Water through Atmosphere in an Optimized Approach for Future Cities in India " Conference On Technologies For Future Cities (CTFC) 2019,(April 2019), SSRN(Elsevier), (March 5, 2019) ISSN: 2278-3075, ISBN: 978-93-82626-27-5, July 2019,pp.1-6

72. Karthik Nagarajan, Mihir Patilhande (2017): "Rehabilitation cost-EffEffectiveuse for Sustainable Rural Development - A Case Study of Landslide affected Dasgaon village in Maharashtra " at International Conference on Recent Trends, Emerging Issues in Science, Engineering, Environment and Technology (ICRESEET-2017 ISSN (Online) 2456-1290, DOI - 01.1617/vol4iss1pid002517 Vol 2, Issue 4, April 2017,pp:25-31, held at Goa from April 8th -9th 2017 (April 2017)

73. Karthik Nagarajan, Rahul Hazarika, Raunak Upadhyay,Sayali Joshi, Tarandeep Singh (2019) : "ASustainable Smart Technique for Generating Water from Atmosphere for Future Cities - A Thermal Electric Cooling Approach," Conference On Technologies For Future Cities (CTFC) (March 5, 2019), SSRN(Elsevier), (Apr 2019) ISSN: 2278-3075 July 2019,ISBN: 978-93-82626-27-5,pp.1-6

74. Karthik Nagarajan, Rahul Hazarika, Raunak Upadhyay, S ayali Joshi, Tarandeep Singh (2019): "A sustainable smart Technique for Generating Water from Atmosphere for Future Cities - A Thermal Electric Cooling Approach", Conference On Technologies For Future Cities (CTFC) (March 5, 2019), SSRN(Elsevier), (Apr 2019) ISSN: 2278-3075 July 2019, ISBN: 978-93-82626-27-5, pp.1-6

75. Karthik Nagarajan, Raju Narwade and Akshata Mestry (2020): "Estimation of Water Balance components of Watersheds in the Manjira River Basin Using Swat Model and G.I.S.", International Journal of Engineering and Advanced Technology (IJEAT), ISSN: 2249- 8958, Volume-9, Issue-3, (February 2020).

76. Karthik Nagarajan, Raju Narwade and Sanika Kandalekar (2019): "Experimental Study on Performance of Portland Based Pervious Concrete using Chemical Supplementaries and Varying Curing Technique " i-manager Journal of Civil Engineering, (Nov-2019), ISSN NO: 2249-7455, Vol:9, Issue:4, pp.21-28.

77. Karthik Nagarajan, Raju Narwade, Ashwini P, Sachin Panhalkar, Vedashree Sameer Kulkarni, Aishwarya Pramod Hingmire (2022): "Employing Digital Elevation Model (DEM) for Floodplain mapping with applications of HEC-HMS, HEC-RAS,

and ArcGIS: A Review Approach ", STM Journals, (July 2022), eISSN: 2230-7990, p-ISSN: 2321-421X, Volume 13, Issue 2, pp-1-16.

78. Karthik Nagarajan, Raju Narwade, Harshal Pathak, Sachin Panhalkar, Vedashree Sameer Kulkarni, Aishwarya Pramod Hingmire (2022): "Review Paper for Floodplain Mapping with Applications of HEC-HMS, HEC-RAS, and ArcGIS Softwares – A Remote Sensing and GIS Approach ", International Research Journal of Engineering and Technology (IRJET), (June 2022), e-ISSN: 2395-0056, p-ISSN: 2395-0072, Volume 9, Issue 6, pp-812-826.

79. Karthik Nagarajan, Raju Narwade, Harshal Pathak, Sachin Panhalkar, Vedashree Sameer Kulkarni, Aishwarya Pramod Hingmire (2022): "Review Paper for Floodplain Mapping with Applications of HEC-HMS, HEC-RAS, and ArcGIS Softwares – A Remote Sensing and GIS Approach ", International Research Journal of Engineering and Technology (IRJET), (June 2022), e-ISSN: 2395-0056, p-ISSN: 2395-0072, Volume 9, Issue 6, pp-812-826.

80. Karthik Nagarajan, Raju Narwade, Mukkawar Pravin Shankarrao1 (2022): "Meticulous Review of Present Trends in Risk Management of Large-Scale Infrastructure Projects", International Journal of Research in Engineering and Science (IJRES), (July 2022), ISSN (Online): 2320-9364, ISSN (Print): 2320-9356, Volume 10, Issue 7, pp-91-95.

81. Karthik Nagarajan, RajuNarwade and Chowdary Mohanlal (2019): "Applications of 4D G.I.S. Model in Construction Management ", International Journal of Innovative Technology and Exploring Engineering (IJITEE), (July-2019) ISSN: 2278-3075, Volume-8 Issue-9, July 2019, pp.2597-2608

82. Karthik Nagarajan, RajuNarwade and Pradhnya Patil (2019): "Effect of Marble Powder and P.V.A. Fibres on the Strength and Microstructure of Engineered Cementitious Composite by using Non-Destructive Test " SSRG International Journal of Civil Engineering (SSRG-IJCE)" - ISSN: 2348 – 8352, (March-2019), Vol 6, Issue 3, pp. 51-57.

83. Karthik Nagarajan, RajuNarwade and Reshma Kamble (2019): "Water Distribution Network Analysis for A Sawale Village Near Rasayani With Application of Epanet Software - A GIS Approach" International Journal of Management, Technology And Engineering (IJMTE), (May-2019), ISSN NO: 2249-7455, Vol:9, Issue:2, pp.3196-3202

84. Karthik Nagarajan, RajuNarwade and Reshma Patil(2019): "Water distribution network analysis for a rural village near Rasayani with the application of hydraulic software -a Geographic Information System (G.I.S.) approach " Conference on Technologies of Future Cities (CTFC, (2019 Available in the SSRN eLibrary: http://ssrn.com/link/2019-CTFC.html Engineering Research Network (EngRN), Proceedings ISBN 978-93-82626-27-5

85. Kelly Brumbelow, Jacob Torres, Seth Guikema, Elizabeth Bristow, and Lufthansa Kanta. (2007), "Virtual Cities for Water Distribution and Infrastructure System Research," 2007 ASCE: World Environmental and Water Resources Congress 2007: Restoring Our Natural Habitat

86. L.U.L.C. of Western Ghat by Comparing N.D.V.I. and N.D.W.I. "Journal of Remote Sensing & G.I.S., (Dec-2019), ISSN: 2469-4134, Vol-08, Issue-04, pp:01-07.

87. Laishram Kanta Singha, Madan K. Jhaa and V. M. Chowdary (2017)," Multi-criteria analysis and G.I.S. modeling for identifying prospective water harvesting and artificial recharge sites for sustainable water supply. "Elsevier, Journal of Cleaner Production,142, (4),1436-1456.

88. Lewis A. Rossman Water Supply and Water Resources Division National Risk Management Research Laboratory Cincinnati, OH 45268 NATIONALS 'E.P.A.N.E.T. 2USERS MANUAL."

89. Loveland, T.R., Merchant, J.W., Brown, J.F., Ohlen, D.O., Reed, B.C., Olson, P., Hutchinson, J., 1995. Seasonal land-cover regions of the United States. Ann. Assoc. Am. Geogr. 85 (2), 339–355.

90. "Manual on Water Supply and Treatment" Third Edition, The G.O.I., Ministry of Urban Development, New Delhi, March 1991, pp 319 "Open Data Kit: Implications for the Use of Smartphone Software Technology for Questionnaire Studies in International Development.",Apr.2014, pp 410-416.

91. M. Günthera, D. Camhya, D. Steffelbauera, M. Neumayerb, D. Fuchs-Hanuscha. (2015), "Showcasing an intelligent water network based on the experimental water distribution system, "Elsevier: Procedia Engineering,119,450 – 457

92. M. P. Suryawanshi, Karthik Nagarajan and Raju Narwade(2021): "Optimisation of Cost in Ground Improvement for Upcoming Navi Mumbai International Airport ", International Journal

of Engineering and Advanced Technology (IJEAT), (Aug-2021) ISSN: 2249-8958 (Online), Volume-10 Issue-06, pp.174-181.

93. M. Polycarpou.(2014)," A Low-Cost Sensor Network for Real-Time Monitoring and Contamination Detection in Drinking Water Distribution Systems, "IEEE Sensors Journal,14(8), 2765- 2772.

94. M. Tamer Ayvaz and Elcin Kentel" Identification of the Best Booster Station Network for a Water Distribution System." Journal of Water Resources Planning and Management (2014).

95. Mahesh S. Singh, Karthik Nagarajan, Raju Narwade, Tejaswini D. N.(2019): "Factors affecting the labor productivity of brickwork and analyzing them using RII method " International Journal of Advanced Technology and Engineering Exploration, (IJATEE), (May-2019), ISSN (Print): 2394-5443 ISSN (Online): 2394-7454, Vol-06, Issue-54, pp.143-151,

96. May, R.J., Dandy, G.C., Maier, H.R., Nixon, J.B., 2008. Application of partial mutual information variable selection to ANN forecasting of water quality in water distribution systems. Environmental Modelling & Software 23 (10e11), 1289-1299.

97. McIver, Friedl, M.A., D.K., Hodges, J.C., Zhang, X.Y., Muchoney, D., Strahler, A.H., Woodcock, C.E., Gopal, S., Schneider, A., Cooper, A., Baccini, A., 2002. Global land Meenu Ramadas and Alok Kumar Samantaray.(2017), "Applications of Remote Sensing and G.I.S. in Water Quality Monitoring and Remediation: A State-of-the-Art Review," Springer, Water Remediation, Energy, Environment and Sustainability,225- 246

98. Michael E. Tryby, Dominic L. Boccelli, James G. Uber and Lewis A. Rossman "Facility Location Model for Booster Disinfection of Water Supply Networks." Journal of water resources planning and management (2002).

99. Mihir Patilhande and Karthik Nagarajan (2017):"Rehabilitation and Cost Effective house for Sustainable Rural Development - A Case Study of Landslide affected Dasgaon village in Maharashtra " at International Conference on Recent Trends, Emerging Issues in Science, Engineering, Environment and Technology (ICRESEET-2017) : ISSN (Online) 2456-1290, DOI - 01.1617/ vol4iss1pid002517,Vol 2, Issue 4, April 2017,pp:25-31, held at Goa from April 8th -9th 2017 (April 2017)

100. Mihir Patilhande, Karthik Nagarajan, Siddhesh Satere, Pulkit Gupta (2017): "Rehabilitation and Cost-Effective house for

Sustainable Rural Development - A Case Study of Landslide affected Dasgaon village in Maharashtra" International Journal of Engineering Research in Mechanical and Civil Engineering (IJERMCE),(April 2017)Vol 2, Issue 4,e-ISSN: 2456-1290, pp.25-31

101. Mit J Kotecha,Karthik Nagarajan, Shruti Kanga, Suraj Kumar Singh, Ritwik Nigam, Achala Shakya (2022): "Analysis of Urban Heat Island Effect in Rajkot City Using Geospatial Techniques", Geospatial Technology for Landscape and Environmental Management, (Jan 2022)(Print ISSN2345-4415; Online ISSN2345-4423), Volume 10(Issue 2), Serial No 26, (Jan 2022), pp.351-370.

102. Mohd. Abbas H. Abdy Sayyed, Rajesh Gupta, Tiku Tanyimboh. (2017), "Flow-Based Penalty In Ga For Optimal Design Of Water Distribution Network, "Hydro-2017 International, L.D.College of Engineering Ahmedabad, India, 1439-1446.

103. Monteiro's, DiasS.T,Coelhi. "Modelling of chlorine decay in drinking water supply using Epanet M.S.X." ELSEVIER (2013).

104. Muhammad Nadeem Sharif. "Optimization of chlorination booster in drinking water distribution network for Al-Khobar city in Saudi Arabia. Saudi Society for Geoscience (2016).

105. N. Khanal ., Buchberger, S. G., Mckenna, S. A., Clark, R. M., and Grayman "Vulnerability assessment of water distribution system to chemical intrusions." World Water and Environmental Resources Congress (2016).

106. Olsson, R.J., Kapelan, Z., Savic, D.A., 2009. Probabilistic building block identification for the optimal design and rehabilitation of water distribution systems. Journal of Hydroinformatics 11 (2), 89-105.

107. Ostfeld, A., Tubaltzev, A., 2008. Ant colony optimization for least-cost design and operation of pumping water distribution systems. Journal of Water Resources Planning and Management-ASCE 134 (2), 107-118.

108. Pallavi Patil, Karthik Nagarajan, RajuNarwade(2019): " Resource Management of Infrastructural Project for Future Cities: A Re Modified Minimum Moment method " International Journal of Management, Technology And Engineering (IJMTE), (Feb-2019), ISSN NO: 2249-7455, Vol:9, Issue:2, pp.1094-1099

109. Panaskar S, Karthik Nagarajan, Raju Narwade(2019): "Analysis of Changes in LULC of Western Ghat by Comparing NDVI and

NDWI " Journal of Remote Sensing & GIS, (Dec-2019), ISSN: 2469-4134, Vol-08, Issue-04, pp:01-07.

110. Pradhnya Patil, Karthik Nagarajan, RajuNarwade(2019): "Effect of Marble Powder and PVA Fibres on the Strength and Microstructure of Engineered Cementitious Composite by using Non-Destructive Test" "SSRG International Journal of Civil Engineering (SSRG-IJCE)" - ISSN: 2348 – 8352, (March-2019), Vol 6, Issue 3, pp.51-57.

111. Pranav Andhyal, Karthik Nagarajan and Raju Narwade (2021): "Applications of 5D CAD for Billing in Construction using G.I.S.", International Journal of Innovative Technology and Exploring Engineering (IJITEE), (May-2021) ISSN: 2278-3075, Volume-10 Issue-07, pp.74-82

112. Priyanka Bhatkar, Raju Narwade and Karthik Nagarajan (2019): "A Case Study on Impact of Tar-Ball Pollution on the Beaches of Alibaug (Maharashtra)" i- manager Journal of Civil Engineering, (Aug- 2019), ISSN NO: 2249-7455, Vol:9, Issue:3, pp.50-58.

113. R.S. Savita, H.K. Mittal, U. Satishkumar, P.K. Singh, K.K. Yadav. (2018), "Delineation of Groundwater Potential Zones Using Remote Sensing and G.I.S. Techniques in Kanakanala Reservoir Subwatershed, Karnataka, India" International Journal of Current Microbiology and Applied Sciences, 7(1),273-288

114. Rafi Schwartz, Nurit Oliker, and Avi Ostfeld "World Environmental and Water Resources Congress" ASCE (2013).

115. Rajnish Singh, Karthik Nagarajan and Raju Narwade (2021): "Analysis of Labour Productivity and Determining the Parameters Which Affects it in Aluminum Formwork System", International Journal of Innovative Technology and Exploring Engineering (IJITEE), (Oct-2021) ISSN: 2278-3075 (Online), Volume-10 (Issue-12), pp.16-23.

116. Ramesh L. Dikpal T. J. Renuka Prasad K. Satish. (2018), "Evaluation of morphometric parameters derived from Cartosat-1DEM using remote sensing and

117. Ramesh L. Dikpal T. J. Renuka Prasad K. Satish. (2018), "Evaluation of morphometric parameters derived from Cartosat-1DEM using remote sensing and Ramos, H., Tamminen, S., Covas, D., 2009. Water supply system performance for different pipe materials part II: sensitivity analysis to pressure variation. Water Resources Management 23 (2), 367e393.

118. Reshma Kamble, Karthik Nagarajan and Raju Narwade (2019): "Water Distribution Network Analysis for A Sawale Village Near Rasayani With Application of Epanet Software - A GIS Approach "International Journal of Management, Technology and Engineering (I.J.M.T.E.), (May-2019), ISSN NO:2249-7455, Vol:9, Issue:2, pp.3196-3202

119. Robert Sitzenfrei, Michael Möderl, Wolfgang Rauch (2013)," Automatic generation of water distribution systems based on G.I.S. data," Elsevier: Environmental Modelling & Software,47,138-147.

120. Rodriguez, M.J., Serodes, J.B., 1999. Assessing empirical linear and non-linear modeling of residual chlorine in urban drinking water systems. Environmental Modelling & Software 14 (1), 93-102.

121. Rolf SprengelRolf SprengelKarl RohrKarl RohrH. Siegfried Stiehl. Siegfried Stiehl (May 1998), "Thin-plate spline approximation for image registration," Engineering in Medicine and Biology Society. Bridging Disciplines for Biomedicine. Proceedings of the 18th Annual International Conference of the IEEEVolume: 3 SourceIEEE XploreDOI: 10.1109/IEMBS.1996.652767

122. Rossman, L.A., 2000. E.P.A.N.E.T. 2 User Manual. National Risk Management Research Laboratory e U.S. Environmental Protection Agency, Cincinnati, Ohio Savic, D.A., Walters, G.A., 1997. Genetic algorithms for least-cost design

123. S.Alvisi, E. Creaco and M. Franchini. (2011), " Segment identification in water distribution systems, "Urban Water Journal, 8(4),203-217

124. S.Chandramouli And P.Malleswararao(2011) "Reliability-Based Optimal Design Of A Water Distribution Network For Municipal Water Supply "International Journal of Engineering and Technology,3(1),13-19

125. Sahil Waquar Khan, Karthik Nagarajan, and Raju Narwade (2022): "Risk assessment and challenges faced in repairs and rehabilitation of dilapidated buildings", Journal of Rehabilitation in Civil Engineering (JRCE), (Jan 2022)(Print ISSN2345-4415; Online ISSN2345-4423), Volume 10(Issue 2), Serial No 26, Spring 2022, pp.93-112.

126. Sanika Kandalekar, Karthik Nagarajan, RajuNarwade(2019): "Feasibility of Pervious Concrete Pavement: A Case Study of Karanjade Node, Panvel " SSRG International Journal of Civil

Engineering (SSRG-IJCE)",(March-2019), ISSN: 2348-8352, Volume 6, Issue 3 - March 2019, pp.35-40.

127. Sanjay Dahasahasra(2016), "Smart Cities in India - hurdles and solutions for 24x7 water supply" Indian Plumbing today, May 2016, pp 14-18

128. Santhosh LG Nitte Meenakshi Institute of Technology "Simulation of Hydraulic Parameters in Water Distribution Network Using E.P.A.N.E.T. and G.I.S."

129. Savic, D.A., Bicik, J., Morley, M.S., 2011. A DSS generator for multiobjective optimization of spreadsheet-based models. Environmental Modelling & Software 26 (5), 551-561.

130. Schaake, J., Lai, D., 1969. Linear Programming and Dynamic Programming Applications to Water Distribution Network Design. No. 116. Department of Civil Engineering, Massachusetts Institute of Technology, U.S.A.

131. Shatri Aditya, Raju Narawde and Karthik Nagarajan (2019): "Integrated Land-Use Zoning, UsingTopographical Data: Optimizing Vacant Space for Urbanization at Akole Taluka, Maharashtra, India "International Journalof Advanced Research in Engineering and Technology (I.J.A.R.E.T.), Volume 10, Issue 1, (Jan-Feb 2019), pp. 188-199.

132. Shobana Jadhav, Karthik Nagarajan and Raju Narwade (2019): "Best Feasible Transportation Route Analysis for Delivering Ready Mixed Concrete (R.M.C.) - A Geographic Information System (G.I.S.) Approach "International Research Journal of Engineering and Technology (I.R.J.E.T.)," (Feb-2019) Vol:6, Issue:2,

133. Shrenik Shah, Karthik Nagarajan, Raju Narwade and (2020): "Assessment Of Urban Utilities For Mumbai City Using 3d Modeling Techniques", International Journal of Civil Engineering and Technology (IJCIET), (May 2020) ISSN Print: 0976-6308 and ISSN Online: 0976-6316, Volume-11, Issue-5.

134. Shweta Panaskar, Sachin Bhere, Karthik Nagarajan, and Raju Narwade(2022): "Validating the Gravity Recovery and Climate Experiment (GRACE) Terrestrial Water Storage Dataset using Terra MODIS NDVI Anomaly ", Water Resource Engineering and Management, (May 2022), ISSN: 2349-4336, Volume 9, Issue 1, PP.19-28.

135. Singh, A., 1989. Review article: digital change detection techniques using remotely sensed data. Int. J. Remote Sens. 10 (6), 989–1003. https://doi.org/10.1080/ 01431168908903939.

136. Sonali Kadu, Karthik Nagarajan and Raju Narwade(2021): "Impacting the Timeline of Railway Project Using Relative Importance Index Method (RII), and Important Index (IMPI) ",Strad Research, (July-2021) ISSN: 0039-2049, Volume-08 Issue-07, pp.138-148.

137. Sou-SenLeu and Quang-NhaBui (2016)," Leak Prediction Model for Water Distribution Networks Created Using a Bayesian Network Learning Approach," Springer, Water-Open Access Journal,30(8), 2719-2733

138. Srila Gopal, Rajiv Sarkar et al. "Study of water supply & sanitation practices in India using geographic information systems: Some design & other considerations in a village setting" Indian J Med Res 129, March 2009

139. Stian Bruaset, Sveinung Sægrov and Rita Ugarelli.(2017), " Performance-based modelling of long-term deterioration to support rehabilitation and Stian Bruaset, Sveinung Sægrov and Rita Ugarelli.(2017), " Performance-based modeling of long-term deterioration to support rehabilitation and investment decisions in drinking water distribution systems " Urban Water Journal,15(1),46-52

140. Stroppiana, D, Boschetti, M., Brivio, P.A., Bocchi, S., 2009. Multi-year monitoring of rice crop phenology through time series analysis of M.O.D.I.S. images. Int. J. Remote Sens. 30 (18), 4643–4662. https://doi.org/10.1080/01431160802632249.

141. Sunilkumar Patel and Karthik Nagarajan (2017):" Sustainable Smart Blue Roof Network System with application of Geographic Information System (G.I.S.) " at International Conference of Emerging technology in Structural Engineering (ETSE-2017): ISSN (Online) 2456-1290,Vol:2, Issue:3, pp.423-428, March 2017,held at K.D.K. College of Engineering, Nagpur.

142. Sunilkumar Patel, Karthik Nagarajan and RajuNarwade (2017): " Modelling Urban Hydrology To Assess The Potential Of Blue Roof Network System " at 22nd International Conference on Hydraulics, Water Resources and Costal Engineering-HYDRO 2017, Dec 2017, held at L.D. College of Engineering, Ahmedabad from21th and 23rd (Dec 2017)

143. Sunilkumar Patel, Karthik Nagarajan and Raju Narwade (2017): "Sustainable Smart Blue Roof Network System with the application of Geographic Information System (G.I.S.), "International Journal of Engineering Research in Mechanical

and Civil Engineering (I.J.E.R.M.C.E.), (March 2017) Vol:2, Issue:3, pp.423-428, ISSN (Online) 2456-1290

144. Tamminen, S., Ramos, H., Covas, D., 2008. Water supply system performance for different pipe materials part I: water quality analysis. Water Resources Management, 22 (11), 1579-1607.

145. Tanvi Nijampurkar, Karthik Nagarajan, Raju Narwade (2020) "Analysis of Sedimentation Deposition in Morbe Reservoir by Using R.S. and G.I.S.", Journal of Remote Sensing & G.I.S., (April 2020), ISSN: 2230-7990 (Online), ISSN 2321-421X (Print), Volume-11, Issue-01, pp-16-24, April 2020. DOI:https://doi.org/10.37591/.v11i1.813

146. Tanvi Nijampurkar, Karthik Nagarajan, Raju Narwade(2020): "Analysis of Sedimentation Deposition in Morbe Reservoirby Using RS and GIS", Journal of Remote Sensing & GIS, (April 2020), ISSN: 2230-7990 (Online), ISSN 2321-421X (Print), Volume-11, Issue-01, pp-16-24, April 2020.

147. Teunis, K. K. Fleming, J. Yang, C. L. Moe, and M. W. Lechevallier "Enteric Virus Infection Risk from Intrusion of Sewage into a Drinking Water Distribution Network" Environment Science and Technology A.C.S. publication (2010).

148. Theofanis P. Lambrou, Christos C. Anastasiou, Christos G. Panayiotou, and Marios M. Polycarpou.(2014)," A Low-Cost Sensor Network for Real-Time Monitoring and Contamination Detection in Drinking Water Distribution Systems, "IEEE Sensors Journal,14(8), 2765- 2772.

149. Thomas Walski and Enrico Creaco2. (2016)" Selection of Pumping Configuration for Closed Water Distribution Systems," American Society of Civil Engineers, Water Resources Planning and Management,142, (6),04016009-1 to 6

150. Todini, E., 2000. Looped water distribution networks design using a resilience index- based heuristic approach. Urban Water 2 (2), 115-122.

151. Torres, J.M., Brumbelow, K., Guikema, S.D., 2009. Risk classification and uncertainty propagation for virtual water distribution systems. Reliability Engineering & System Safety 94 (8), 1259-1273.

152. Uzair (Sam) Shamsi (2004), "G.I.S. Applications for Water Distribution Systems," *Journal of Water Management Modeling* R220-21. DOI: 10.14796/JWMM.R220-21.

153. Vairavamoorthy, K., and Ali, M. "Least Cost Design of Water Distribution Networks" Proceeding of the First International Conference on New I.T. for

154. Vidya S. Gavekar & V. D. Nandavdekar.(2013), "Mapping Of Water Distribution Network Through Geographical Information System (G.I.S.)." International Journal of Civil Engineering,2(5),33-44.

155. Viovy, N., Arino, O., Belward, A.S., 1992. The Best Index Slope Extraction (BISE): a method for reducing noise in N.D.V.I. time-series. Int. J. Remote Sens. 13 (8), 1585– 1590. https://doi.org/10.1080/01431169208904212.

156. Vishal Ambad, Karthik Nagarajan and Raju Narwade(2021): "Impact of controlled permeable formwork liner against chloride penetration on the concrete structures ",Journal of Research on Engineering Structures and Materials (JRESM), (Nov 2021) (p-ISSN: 2148-9807; o-ISSN: 2149-4088), Volume 8(Issue 1), pp.117-126.

157. Vivek Kumar, Raju Narwade and Karthik Nagarajan (2019): "Determination of Water Budget of the Lower Godavari River Basin Using G.I.S. and SWAT "International Journal for Research in Engineering Application & Management (I.J.R.E.A.M.), (Oct-2019), ISSN: 2454-9150, Vol-05, Issue-07, pp.41-47.

158. Waldo Ojeda-Bustamante, Juan Manuel Gonza´ lez-Camacho, Ernesto Sifuentes- Ibarra, Esteban Isidro, Luis Rendo´n-Pimentel. (2007),"Using spatial information systems to improve water management in Mexico " Science direct: Agricultural Water management, 81-88.

159. Walski, T.M., Downey Brill, E., Gessler, J., Goulter, I.C., Jeppson, R.M., Lansey, K., Lee, H.-L., Liebman, J.C., Mays, L., Morgan, D.R., Ormsbee, L., 1987. Battle of the network models: epilogue. Journal of Water Resources Planning and Management 113 (2), 191-203.

160. Walters, G.A., Halhal, D., Savic, D., Ouazar, D., 1999. Improved design of "Anytown" distribution network using structured messy genetic algorithms. UrbanWater 1 (1), 23-38.

161. Wang, Q.J., 2008. A Bayesian method for multi-site stochastic data generation: dealing with non-concurrent and missing data, variable transformation, and parameter uncertainty. Environmental Modelling & Software 23 (4), 412-421.

162. World Urbanization Prospects, United Nations, (2010)

163. Wu, J., Lu, J., Wang, J., 2009a. Application of chaos and fractal models to water quality time series prediction. Environmental Modelling & Software 24 (5), 632-636.

164. Wulder, M.A., Masek, J.G., Cohen, W.B., Loveland, T.R., Woodcock, C.E., 2012. Opening the archive: how free data has enabled the science and monitoring promise of Landsat. Remote Sens. Environ. 122, 2–10. https://doi.org/10.1016/j.rse.2012.01. 010.

165. Zhang, W., Miller, C. T., and Digiano, F. A. "Bacterial regrowth model for water distribution systems incorporating alternating split-operator solution technique." Jour. Environmental Engineering, (2015).

WEB REFERENCES

- https://cfpub.epa.gov(EPA Science Inventory)
- https://bhuvan.nrsc.gov.in/bhuvan_links.php. (Bhuvan)
- http://www.lib.utexas.edu/maps/ams/india(Topo sheet)
- http://www.qgis.org(QGIS)
- http://epanet.de/ghydraulics/index.html.en(GHydraulics: Plug-in)
- https://plugins.qgis.org/plugins/Qgis2threejs/(Qgis2threejs: Plug-in)
- www.waterworld.com(water world)
- www.waterline.com
- www.springer.com
- www.researchgate.net
- www.ascelibrary.org
- https://www.epa.gov/water-research/epanet(Epanet).
- www.elsevier.com
- www.sciencedirect.com
- www.nptel.ac.in
- www.researchgate.net
- www.youtube.com
- www.wikipedia.com
- www.kobotoolbox.com
- http://gyaninformationpedia.blogspot.com

- https://earth.esa.int/web/eoportal/satellite-missions/i/irs-p6
- https://www.isro.gov.in/Spacecraft/cartosat-1
- https://pgbooks.ru/archive/blog/geoblog/1510/
- www.microsoft.com/windows/directx/

BOOK

- Photogrammetry, G.I.S. & Remote Sensing- S S Manugula, Veeranna Bommakanti (2020)
- George Joseph. (2018). "Fundamentals of Remote Sensing" 3rd Edition, Universities Press, Hyderabad, India.
- Thomas Lillesand, Ralph W. Kiefer, Jonathan Chipman (2015). "Remote Sensing and Image Interpretation Edition-7th Edition", Wiley, The United States of America
- Pramod R. Bhave, Rajesh Gupta (2006)."Analysis of Water Distribution Networks" Alpha Science International, Oxford, U.K.

Publications

PAPER PUBLISHED

“Computational based simulation considering various pipe materials in design and analysis of water distribution networks” at 22nd International Conference on Hydraulics, Water Resources, and Coastal Engineering- HYDRO 2017, Dec 2017, held at L.D. College of Engineering, Ahmedabad from21th and 23rd (Dec 2017).

“Application of Geographic information system for water distribution networks through Quantum G.I.S. plug-in with hydraulic simulation for infrastructure and development planning” The 38th Asian Conference of Remote Sensing A.C.R.S., (Oct 2017), New Delhi.

“Water Distribution Network: A Remote Sensing and G.I.S. Approach” at Next Frontiers In Civil Engineering: Sustainable and Resilient Infrastructure, held at I.I.T. from November 30 to December 01 2018.

Awards and Recognition

First Rank for the Final Round (Maharashtra State Level) Avishkar 2019-20 of 14th Maharashtra State Inter-University Avishkar Research Convention: 2019-20, with Research project title: "Novel Application of Open-Source G.I.A. data for Sustainable Water Management - A Case Study using Remote Sensing" under Engineering and Technology discipline in Teachers Category held at Kalina Campus, the University of Mumbai held on January 28-31, 2020

Silver Medal for the Final Round (University Level) of 14th Inter-Collegiate / Institute / Department Avishkar Research Convention: 2019-20 held at Fr.C. Rodrigues Institute of Technology, Vashi, Navi Mumbai on January 02, 2020

Selected for the Final Round (Zonal Level) of 14th Inter-Collegiate / Institute / Department Avishkar Research Convention: 2019-20 held at Ramrao Adik Institute of Technology, Nerul, Navi Mumbai on December 21, 2019, for all Engineering Colleges

Winner at Final Round of **IITB-ISRO-AICTE MAPATHON**: A national-level collaborative Indian Mapping with the topic as "Sustainable Water Transmission Network Design- A Remote Sensing and Geographic Information System Approach, Feb 2021

Best Network Institute Coordinator Award (IIRS ISRO Outreach Program) for Contribution towards promoting Geospatial Technology and its applications in the Year 2018 Indian Institute of Remote Sensing**, Indian Space Research Organization (I.S.R.O.).**

Secretary, Indian Society of Remote Sensing (I.S.R.S.) MUMBAI CHAPTER Tenure 2019-2021 Pillai H.O.C. College of Engineering & Technology, Rasayani. Maharashtra

First Rank for the Final Round (Maharashtra State Level) Avishkar 2019-20 of 14th Maharashtra State Inter-University Avishkar Research Convention: 2019-20, with Research project title: “Novel Application of Open-Source G.I.A. data for Sustainable Water Management - A Case Study using Remote Sensing” under Engineering and Technology discipline in Teachers Category held at Kalina Campus, the University of Mumbai held on January 28-31, 2020

(L to R): **Dr. Sunil Patil** (Organizing Secretary & Director, Department of Students Development UoM), **Dr. Ajay Deshmukh** (Registrar, UoM), **Mr. Karthik N** (Asso.Prof, PHCET), **Dr. Ravindra Kulkarni** (Hon'ble Pro Vice Chancellor, UoM), **Dr. Vinod Patil** (Director, Board of Examinations and Evaluation, UoM)

हॅटट्रिकचा 'आविष्कार'... मुंबई विद्यापीठाच्या कलिना कॅम्पसमध्ये पार पडलेल्या १४व्या आंतरविद्यापीठ 'आविष्कार' संशोधन स्पर्धेत मुंबई विद्यापीठानं नेत्रदीपक कामगिरी केली. एकूण ९३ गुणांसह मुंबई विद्यापीठानं अंतिम विजेतेपद पटकावलं. यावेळी विद्यापीठानं १५ सुवर्ण आणि ६ रौप्य पदकांची घसघशीत कमाई करत आविष्कार संशोधन स्पर्धेवर सलग तिसऱ्या वर्षी विजयी मोहोर उमटवली. यामध्ये मानव्यविद्या, भाषा आणि ललित कला, वाणिज्य, व्यवस्थापन आणि विधी, मूलभूत शास्त्रं, शेती व पशुसंवर्धन, अभियांत्रिकी आणि तंत्रज्ञान, वैद्यकशास्त्र व औषधशास्त्र या विभागांमध्ये विद्यापीठातर्फे एकूण ४८ संशोधक सहभागी झाले होते. या विजेत्यांचं कौतुक करण्यासाठी कुलगुरू डॉ. सुहास पेडणेकर, प्र-कुलगुरू डॉ. रवींद्र कुलकर्णी आणि विद्यार्थी विकास मंडळाचे संचालक डॉ. सुनिल पाटील उपस्थित होते. पदकविजेत्या तरुण संशोधकांविषयी...५

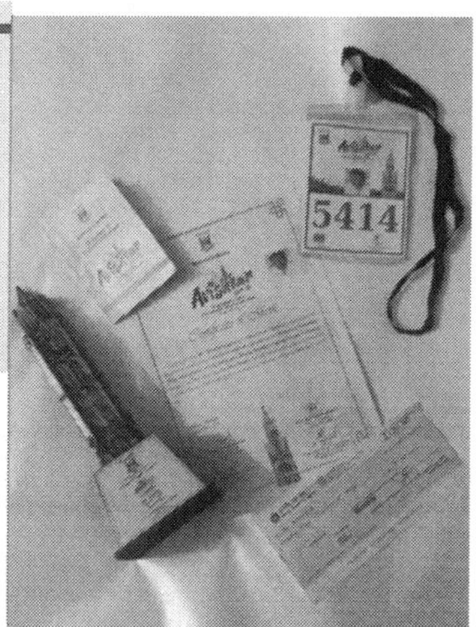

University of Mumbai

Avishkar 2019-20

14th Maharashtra State Inter-University Research Convention

January 28-31, 2020

Certificate of Merit

This is to certify that **Mr. - Karthik Nagarajan** of **University of Mumbai** has participated and presented a research project titled **Novel application of open source GIS data for Sustainable Water Management - A Case Study using Remote Sensing** in **Engineering and Technology** category and **TII** level and secured **First Rank** at the 14th Maharashtra State Inter-University Avishkar Research Convention: 2019-20 organized by University of Mumbai on January 28-31, 2020.

Kalina, Mumbai
January 31, 2020

Prof. Suhas Pednekar
Vice-Chancellor,
University of Mumbai

Prof. Ravindra Kulkarni
Pro Vice-Chancellor,
University of Mumbai

Dr. Ajay Deshmukh
Registrar,
University of Mumbai

Dr. Sunil Patil
Director,
Department of Students' Development,
University of Mumbai and
Organizing Secretary

Silver Medal for the Final Round (University Level) of 14th Inter-Collegiate / Institute / Department Avishkar Research Convention: 2019-20 held at Fr.C. Rodrigues Institute of Technology, Vashi, Navi Mumbai on January 02, 2020

University of Mumbai

14th Inter-Collegiate/ Institute/Department

AVISHKAR

Research Convention: 2019-20

(University Level Research Project Competition)

Certificate of Merit

This is to Certify that **Mr. Nagarajan Karthik** of **Pillais Hoc College of Engineering and Technology, Rasayani** Participated and Presented a Research Project Titled **Parametric Investigation for Sustainable Water Supply Network, Quality and Security Management with Application of GIS and Remote Sensing** in **Engineering and Technology** Category and **TH** Level and Secured **SECOND** Rank at the **Final Round (University Level)** of **14th Inter-Collegiate / Institute / Department Avishkar Research Convention: 2019-20** held at **Fr. C. Rodrigues Institute of Technology, Vashi, Navi Mumbai** on **January 2, 2020**

Dr. (Mrs.) Minakshi Gurav
OSD
Avishkar Research Convention,
University of Mumbai

Dr. Sunil Patil
Director,
Department of Students' Development,
University of Mumbai

January 2, 2020
Vashi, Navi Mumbai

Selected for the Final Round (Zonal Level) of 14th Inter-Collegiate / Institute / Department Avishkar Research Convention: 2019-20 held at Ramrao Adik Institute of Technology, Nerul, Navi Mumbai on December 21, 2019, for all Engineering Colleges

University of Mumbai

14th Inter-Collegiate/ Institute/Department

AVISHKAR

Research Convention: 2019-20

(District/Zonal Level Research Project Competition)

Certificate of Merit

This is to Certify that **Mr. Nagarajan Karthik** of **Pillais Hoc College of Engineering and Technology, Rasayani** Participated and Presented a Research Project Titled **Parametric Investigation for Sustainable Water Supply Network, Quality and Security Management with Application of GIS and Remote Sensing** in **Engineering and Technology** Category and **TH** Level at the **Selection Round** of **14th Inter-Collegiate / Institute / Department Avishkar Research Convention: 2019-20** held at Ramrao Adik Institute of Technology, Nerul, Navi Mumbai on **December 21, 2019** for **All Engineering Colleges of all Districts** zone. The said project is selected for **Final Round** of **14th Inter-Collegiate/Institute/Department Avishkar Research Convention: 2019-20.**

DR. (MRS.) MINAKSHI GURAV
OSD
Avishkar Research Convention
University of Mumbai

Dr. Sunil Patil
DIRECTOR
Department of Students' Development,
University of Mumbai

Place: Nerul, Navi Mumbai
Date: December 21, 2019

Winner at Final Round of **IITB-ISRO-AICTE MAPATHON**: A national-level collaborative Indian Mapping with the topic "Sustainable Water Transmission Network Design- Remote Sensing and Geographic Information System Approach, Feb 2021

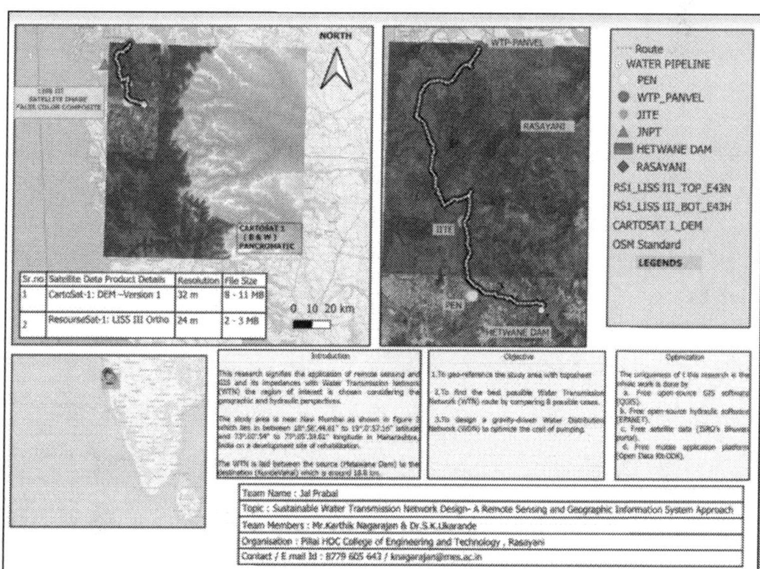

Best Network Institute Coordinator Award (IIRS ISRO Outreach Program) for Contribution towards promoting Geospatial Technology and its applications in the Year 2018 Indian Institute of Remote Sensing**, Indian Space Research Organization (I.S.R.O.).**

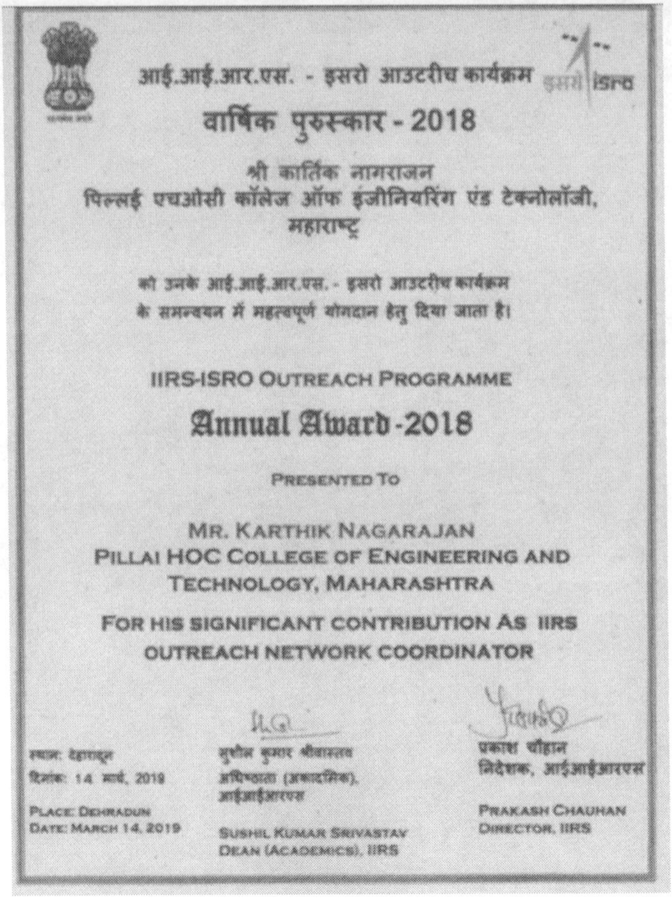

आई.आई.आर.एस. - इसरो आउटरीच कार्यक्रम

वार्षिक पुरस्कार - 2018

श्री कार्तिक नागराजन
पिल्लई एचओसी कॉलेज ऑफ इंजीनियरिंग एंड टेक्नोलॉजी,
महाराष्ट्र

को उनके आई.आई.आर.एस. - इसरो आउटरीच कार्यक्रम
के समन्वयन में महत्वपूर्ण योगदान हेतु दिया जाता है।

IIRS-ISRO OUTREACH PROGRAMME

Annual Award - 2018

PRESENTED TO

MR. KARTHIK NAGARAJAN
PILLAI HOC COLLEGE OF ENGINEERING AND TECHNOLOGY, MAHARASHTRA

FOR HIS SIGNIFICANT CONTRIBUTION AS IIRS OUTREACH NETWORK COORDINATOR

स्थान: देहरादून
दिनांक: 14 मार्च, 2019

PLACE: DEHRADUN
DATE: MARCH 14, 2019

सुशील कुमार श्रीवास्तव
अधिष्ठाता (अकादमिक),
आईआईआरएस

SUSHIL KUMAR SRIVASTAV
DEAN (ACADEMICS), IIRS

प्रकाश चौहान
निदेशक, आईआईआरएस

PRAKASH CHAUHAN
DIRECTOR, IIRS

IIRS-ISRO OUTREACH PROGRAMME

Certificate of Appreciation

PRESENTED TO

PROF. KARTHIK NAGARAJAN
FROM
PILLAI HOC COLLEGE OF ENGINEERING AND TECHNOLOGY

FOR HIS/HER CONTRIBUTION TOWARDS PROMOTING GEOSPATIAL TECHNOLOGY AND ITS APPLICATIONS IN YEAR 2018

THROUGH

IIRS-ISRO OUTREACH PROGRAMME AS A COORDINATOR OF NETWORK INSTITUTE

S.K.SRIVASTAV, DEAN (A), IIRS
PRAKASH CHAUHAN, DIRECTOR, IIRS

DATE: 15 MARCH 2019
PLACE: DEHRADUN

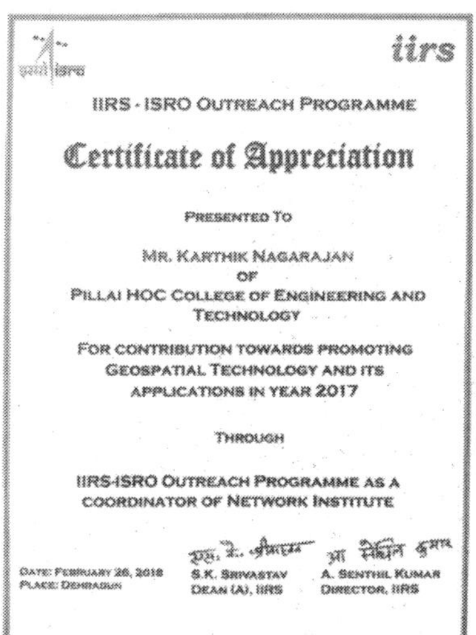

iirs

IIRS - ISRO OUTREACH PROGRAMME

Certificate of Appreciation

PRESENTED TO

MR. KARTHIK NAGARAJAN
OF
PILLAI HOC COLLEGE OF ENGINEERING AND TECHNOLOGY

FOR CONTRIBUTION TOWARDS PROMOTING GEOSPATIAL TECHNOLOGY AND ITS APPLICATIONS IN YEAR 2017

THROUGH

IIRS-ISRO OUTREACH PROGRAMME AS A COORDINATOR OF NETWORK INSTITUTE

DATE: FEBRUARY 26, 2018
PLACE: DEHRADUN

S.K. SRIVASTAV, DEAN (A), IIRS
A. SENTHIL KUMAR, DIRECTOR, IIRS

IIRS-ISRO OUTREACH PROGRAMME

Certificate of Appreciation

PRESENTED TO

PROF. KARTHIK NAGARAJAN
FROM
PILLAI HOC COLLEGE OF ENGINEERING & TECHNOLOGY, MAHARASHTRA

FOR HIS/HER CONTRIBUTION TOWARDS PROMOTING GEOSPATIAL TECHNOLOGY AND ITS APPLICATIONS IN YEAR 2019

THROUGH

IIRS-ISRO OUTREACH PROGRAMME AS A COORDINATOR OF NETWORK INSTITUTE

S.K.SRIVASTAV, DEAN (A), IIRS
PRAKASH CHAUHAN, DIRECTOR, IIRS

DATE: MARCH 04, 2020
PLACE: DEHRADUN

Secretary, Indian Society of Remote Sensing (I.S.R.S.) MUMBAI CHAPTER Tenure 2019-2021 Pillai H.O.C. College of Engineering & Technology, Rasayani. Maharashtra

Indian Society of Remote Sensing
Serving Humanity through Space Technology
ISRS

Indian Society of Remote Sensing (ISRS) was established in 1969 with the main objective of advancement and dissemination of remote sensing technology in the fields of mapping, planning and management of natural resources and environment by organising seminars/symposia and by publishing a quarterly journal (JISRS), bulletins, proceedings, etc.

Starting with 56 members, the Society has now grown into a premier professional body with membership of over **5286** members. Currently, ISRS is a member of the International Society of Photogrammetry and Remote Sensing (ISPRS).

More Details
https://www.isrs-india.org/

Mr. Karthik Nagarajan
Secretary

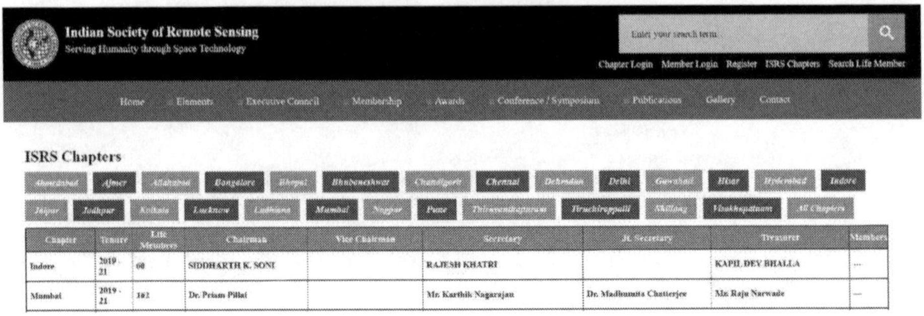

Chapter	Tenure	Life Members	Chairman	Vice Chairman	Secretary	Jt. Secretary	Treasurer	Members
Indore	2019 - 21	60	SIDDHARTH K. SONI		RAJESH KHATRI		KAPIL DEV BHALLA	---
Mumbai	2019 - 21	102	Dr. Priam Pillai		Mr. Karthik Nagarajan	Dr. Madhumita Chatterjee	Mr. Raju Narwade	---

SPECIAL TRAINING PROGRAM FOR UNIVERSITY POSTGRADUATE TEACHERS (ISRO/DOS – N.N.R.M.S. SPONSORED PROGRAM)

The National Natural Resources Management System (N.N.R.M.S.), a national level inter-agency system set-up in 1983 and chaired by the Member, Planning Commission, Govt. of India, steers the E.O. program in the country.

https://www.iirs.gov.in/nnrmssponsoredcourses

Indian Institute of Remote Sensing
Indian Space Research Organisation
Department of Space, Government of India

Enrolment No. 10665

Certificate

This is to certify that **Mr. Karthik Nagarajan** from **Pillai HOC College of Engineering and Technology, Rasayani, Raigad, Maharashtra** has been awarded this certificate on having completed the ISRO-sponsored NNRMS Course on **"Remote Sensing and Geographical Information System in Water Resources"** conducted by this Institute during May 01, 2017 to June 23, 2017.

Course Director | Group Head, PPEG | Dean, Academics | Director, IIRS

Date: June 23, 2017
Place: Dehradun, India

भारत सरकार
अंतरिक्ष विभाग
भारतीय सुदूर संवेदन संस्थान
4, कालीदास मार्ग, पो. बाक्स सं. 135
देहरादून – 248 001, भारत
दूरभाष : +91-135-2524399
फैक्स : +91-135-2741987, 2748041

Government of India
Department of Space
Indian Institute of Remote Sensing
4, Kalidas Road, P.B. No. 135,
Dehradun - 248 001, India
Telephone : +91-135-2524399
Fax : +91-135-2741987, 2748041

CERTIFICATE

This is to certify that the project work entitled **"Water Distribution Network Utility Management Using GeoSpatial Techniques and Hydraulic Simulation Approach"** submitted by **Mr. Karthik Nagarajan** partial fulfilment of the requirements for the successful completion of **NNRMS 2017**. The project presented here in this report is an original work of the candidate and has been carried out in Water Resources Department, Indian Institute of Remote Sensing, Dehradun, under the guidance of **Dr. Praveen K. Thakur**, Scientist SF, IIRS, Water Resources Department and **Mr. Kamal Pandey**, Scientist SD, Geoweb Services, IT & Distance Learning, Indian Institute of Remote Sensing, Dehradun, India.

Praveen 22-6-17
(Dr. Praveen K. Thakur)
Scientist/Engineer 'SF'
Water Resources Department

Pandey 22/6/17
(Mr. Kamal Pandey)
Scientist/Engineer 'SD'
GeoWeb, IT and Distance Learning

SPA
(Dr. S. P. Aggarwal)

Scientist /Engineer 'SG' & Head,
Water Resources Department

INTERNATIONAL TRAINING COURSE "8TH SMALL SATELLITE MISSION"

Centre for Space Science and Technology Education in the Asia Pacific (C.S.S.T.E.A.P.)-Affiliated to United Nations - U.N. at I.I.R.S. (I.S.R.O.) Dehradun from Nov 25 -06th Dec 2019

Centre for Space Science and Technology Education in Asia and the Pacific
(Affiliated to the United Nations)

email : csstеap@iirs.gov.in
csstеapun@gmail.com
website : www.cssteap.org

No. CSSTEAP/8th SSM/2019

Dated: October 18, 2019

Dr. Madhumita Chaterjee
Principal,
Pillai HOC College of Engineering & Technology, Raigad,
Via Panel, Rasayani, Taluka-Khalapur, Dist – Raigad 410207,
Maharashtra, India
Phone-02152252005
Email- principal.phcet@mes.ac.in

Sub: Admission Letter- CSSTEAP International Training Course on 8th Small Satellite Mission at IIRS, (November 25 - December 06, 2019)

Dear Dr. Madhumita Chaterjee,

I am glad to inform you that **Mr. Karthik Nagarajan** from your organization has been selected for admission for the 8th International training course on Small Satellite Mission at IIRS, Dehradun **subject to the condition that all training related expenses including travel from his place to Dehradun and back, their living expenses etc. will be borne by the candidate/organization and submission of duly filled enclosed undertaking form to the Centre.**

1. The candidate is expected to reach Dehradun, India preferably one day in advance i.e. on **November 24, 2019.**
2. Single room furnished accommodation with a common kitchenette in the Hostel will be provided to the candidate, on payment basis at a nominal charge of Indian rupees 100/- per day. However, Indian food is also available on payment basis (Indian rupees 3,000/- per month approximate) in mess at IIRS.
3. The course fee of Rs. 15000/- may please be sent by Demand Draft in favour of Accounts Officer, CSSTEAP, Dehradun or by electronic transfer as per the bank details attached.

The candidate should bring 06 copies of the recent passport size photograph. Candidate is also advised to send medical fitness certificate if not sent before.

The candidate may be advised to confirm his acceptance of joining the course to the Centre by **October 20, 2019** through email (cssteap-admissions@iirs.gov.in, cssteap@iirs.gov.in) or fax (0135 2740785). In case, we do not receive written confirmation either (on email or fax) by **October 20, 2019** (COB), we will be obliged to operate the waiting list.

With regards,

Yours sincerely,
(Dr. S.P. Aggarwal)
Programme Coordinator, CSSTEAP

Copy to:

Mr. Karthik Nagarajan
Associate Professor
Mahatma Education Society, Pillai HOC College of Engineering & Technology,
Raigad, Via Panel, Rasayani, Taluka-Khalapur, Dist – Raigad 410207, Maharashtra, India
Phone: (O) 02/91-252005/250066
(M) 8779605643
Email- knagarajan@mes.ac.in

Dr. A. Senthil Kumar
Director, CSSTEAP

Dr. Prakash Chauhan
Director, IIRS

...quarters	Delhi Office	IIRS Campus	SAC Campus	PRL Campus	URSC Campus
...ampus,	Department of Space	Indian Institute of	Space Application	Physical Research	U R Rao Satellite
...o. 222,	Lok Nayak Bhawan	Remote	Centre	Laboratory	Center
...das Road	3rd Floor, Khan Market	Sensing (ISRO)	Ambawadi Vistar	Navrangpura	Vimanpura Post
...dun – 248001	New Delhi – 110003	No. 4, Kalidas Road	PO. Jodhpur Tekra	Ahmedabad – 380009	Bengaluru – 560017
...1-135-2740737	Tel: 91-11-24694745	Dehradun – 248001	Ahmedabad – 380015	Tel: 91-79-26308980	Tel 91-80-25205252
2740787	Fax: 91-11-2463871	Tel: 91-135-2744583	Tel: 91-79-26913344	Fax: 91-79-263080374	Fax 91-80-25205251
...1-135-2740785		Fax 91-135-2741987	Fax: 91-79-26915843		

Valedictory Function
Conducted by:
and
Dept. of Space , Govt. of India
at
CSSTEAP

iirs
Function
at

MINOR RESEARCH PROJECTS

- Grants Sanctioned as Principal Investigator for Minor research from the University of Mumbai in the Financial Year2019-20

  Minor Research Project No: 810

  Ref No: APD/ICD/2020-21/80 dated September 15 2020,

  Title: "Simulation of 3D Digital Elevation Model (D.E.M.) for Rural Villages by acquiring I.S.R.O. Satellite Image - A Remote Sensing and G.I.S. Approach"

  Amount Sanctioned: Rs. 40,000 /-

- Grants Sanctioned as Principal Investigator for Minor research from the University of Mumbai in the Financial Year2016-17

  Minor Research Project No: 447

  Ref No: APD / 237 / 16/2017

  Title: "Simulation of Water Distribution network (Including Grey Water) with the application of Hydraulic Modelling& G.I.S. software- A GIS and Remote Sensing Approach "

  Amount Sanctioned: Rs. 25,000 /-

RESEARCH COURSES ATTENDED

Sr. No.	Index	Numbers
1	Short Term Training Courses	07
2	Faculty Development Programmes	15
3	Workshops	20
4	Webinars	29
5	Research Grants Received MRPs	2
6	IIRS ISRO Outreach Program conducted	72

MEMBERSHIP OF PROFESSIONAL ORGANISATIONS

No	Membership number		Organization	Year	Registered at
1	I.S.T.E.	LM-86935	Indian society of Technical Education	2012_12_31	IIT (Delhi), campus
2	ISH	LM- 1265	The Indian Society for Hydraulics	2016_12_10	Pune, India
3	ISRS	L-4838	Indian Society of Remote Sensing	2017-07-13	Mumbai, India
4	IAHS	17213	International Association of Hydrological Sciences.	2019-01- 31	New York University, U.S.A.
5	IWWA	LM-9594	Indian Water Works Association	2019-01-31	Mumbai, India
6	I.A.H.R.	64892	International Association of Hydro-Environmental Engineering and Research	2019-04-16	Beijing, China
7	I.S.P.R.S.		International Society of Photogrammetry and Remote Sensing	2019-12-11	Germany
8	ISG	L-2225	Indian Society of Geomatics	2020-01-03	Ahmedabad
9	LCCF		Low Carbon Construct Forum	2022-01-04	Mumbai

POSTER PRESENTATION

Final Round (Maharashtra State Level) Avishkar 2019-20

14th Maharashtra State Inter-University Avishkar Research Convention: 2019-20, Research project title: "Novel Application of Open-Source G.I.A. data for Sustainable Water Management - A Case Study using Remote Sensing" Engineering and Technology discipline in Teachers Category Kalina Campus, the University of Mumbai, January 28-31, 2020

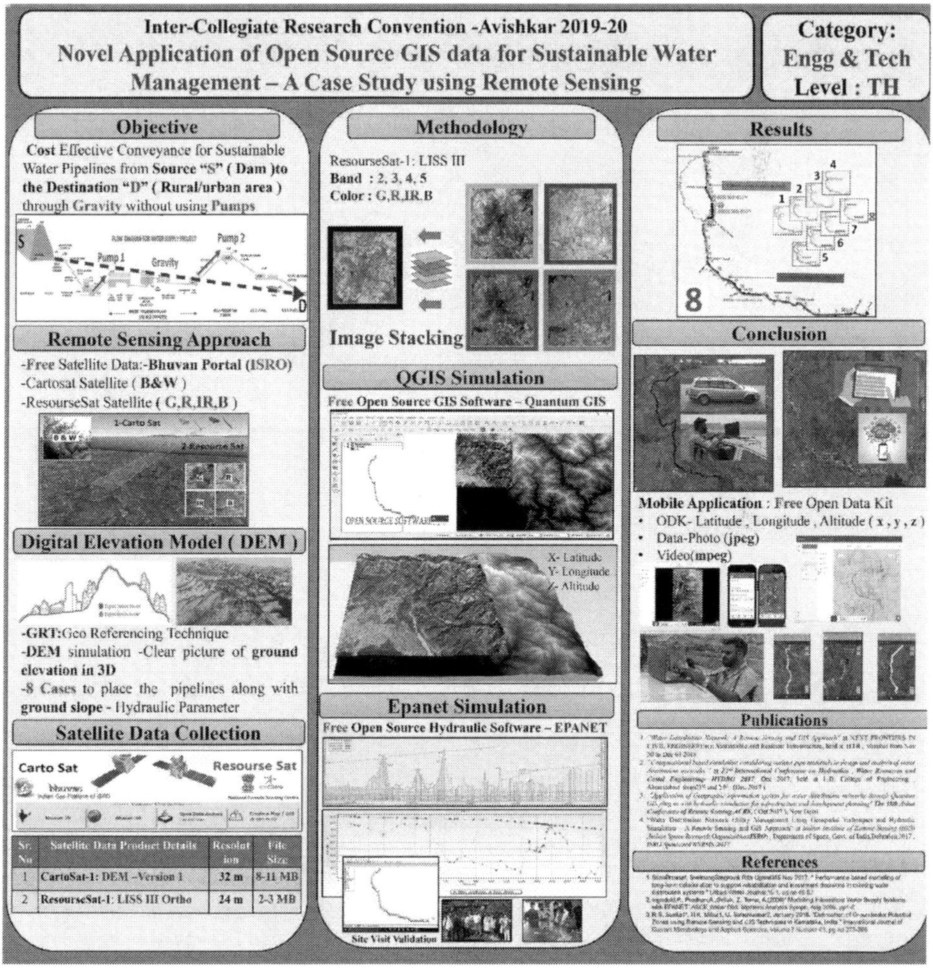

FINAL ROUND (UNIVERSITY LEVEL)

14th Inter-Collegiate / Institute / Department Avishkar Research Convention: 2019-20 Research project title: "Parametric Investigation for Sustainable Water Supply Network, Quality & Security Management with Application of G.I.S. & Remote Sensing ." Engineering and Technology discipline in Teachers Category Fr.C. Rodrigues Institute of Technology, Vashi, Navi Mumbai January 02, 2020

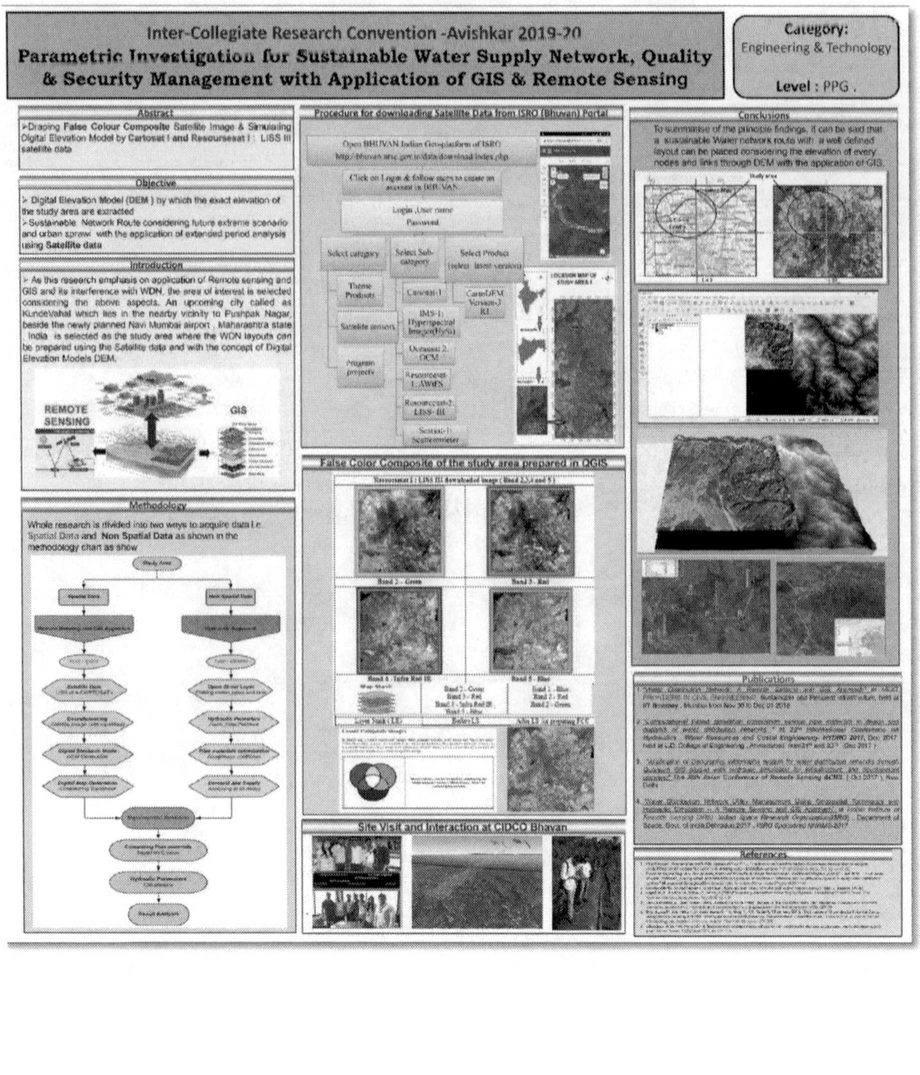

ASIAN CONFERENCE OF REMOTE SENSING - A.C.R.S. 2017

38th Asian Conference on Remote Sensing October 23 - 27, 2017 | Hotel The Ashok, New Delhi, India

Abstract ID- 1311

Application of Geographic Information System for water Distribution Networks through Quantum GIS Plug-in with Hydraulic Simulation for Infrastructure and Development Planning

[1]Karthik Nagarajan , [2]Shrikant Charhate
Department of Civil Engineering ,Pillai HOC College of Engineering and Technology , Rasayani ,Maharashtra , India

Introduction

- Globally the vital renewable natural resource at present and for the future scenario is considered to be **water** and has been striving to address on many research aspects by which we can sustain this resources considering various aspects such as **quality, optimization, maintenance** etc.
- This study combines application of GIS as a framework for managing and integrating data by Quantum GIS i.e(QGIS) with **3 hydraulic plug-in i.e. Ghydraulics , Open layer and iii) Qgis2threejs** later analyzing through hydraulic simulation software in **EPANET**.
- The main challenge faced is the decisions to ensure delivery of water to several locations with optimally sized compatible with existing infrastructure considering the demand supply which later became **the motivating factor for this research work**.

Objectives

To analyze and optimize the water distribution network (WDN) in the most efficient and effective way

- To determine various **hydraulic parameters** like discharge, velocity, pressure and head loss in the water distribution network.
- To compare various **pipe materials** and check their suitability in the system.
- To carry out **GIS implementation** in a given WDN

Study Area and Data Collection

- The water distribution network will be solved using the **Hazen William's formulae (Ref: Table 1)** as it is the best-suited method for solving the gravity based water distribution networks. Hydraulic software EPANET will be used for determining discharge, velocity, head loss, etc. for given WDN.

Table 1 : FORMULA SELECTION

	Parameters for formula selection	Darcy Weisbach	Hazen William	Manning
1	[illegible]	[illegible]	[illegible]	[illegible]
2	[illegible]	[illegible]	[illegible]	[illegible]
3	Flow type	[illegible]	[illegible]	[illegible]
4	Open / Closed	[illegible]	[illegible]	[illegible]
5	[illegible]	Not applicable	Applicable	Not applicable

- **Study area** is an upcoming city near the vicinity of Navi Mumbai airport (Kunde Vahal) . The area is on the verge of becoming a prime location in **midway between Mumbai Pune Corridor.** As the area is in planning stage, there are no well designed WDN. **Fig. 1** shows Toposheet and LISS III **(Ref: Table 2)**

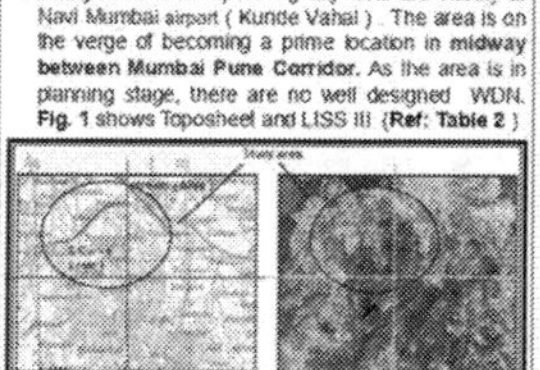

Fig 1: a) STUDY Area b) RESOURCESAT 2,LISS III IMAGE

All 4 bands are stacked togetherin **ERDAS** which helped to understand the topography and other spatial parameters helpful for digitization.

FIG 2: LAYER STACKING

Table 2:DESCRIPTION OF RESOURCESAT-1& LISS-3

[illegible]		
[illegible]	[illegible]	Naming Convention
[illegible]	[illegible]	[illegible]
	Projection Related : [illegible]	
Projection	Datum	Resolution
[illegible]	WGS 84	[illegible]
	Image Related : [illegible]	
[illegible]	Number of Bands	[illegible]
[illegible]	4 (Band 2,3,4,5)	[illegible]

Source: [illegible]

Methodology

- **Single period analysis,(EPANET)** WDN Simulation is and then it runs for that instant only **(Fig.4)** System basically checks all the parameters and determines whether the systemwill be able to meet the water demands or not
- After running the simulation various diameters and flow variation can be seen as in **Fig.3**

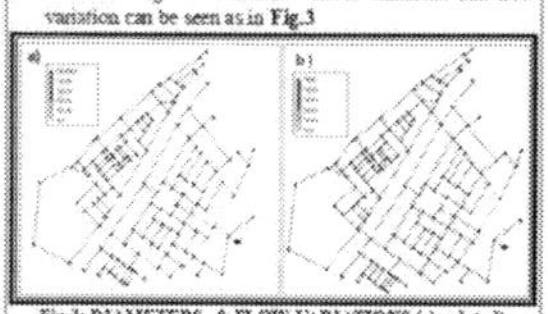

Fig.3: DIAMETERS & FLOW VARIATIONS (simulated)

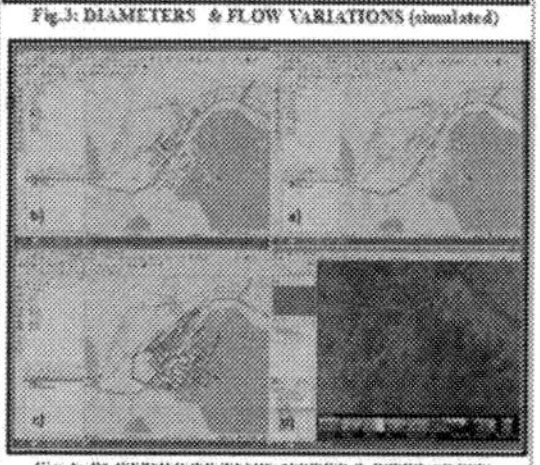

Fig 4: PLOTTING OF TANK ,NODES & PIPES (QGIS)

a) Adding a Tank.

b) Plotting nodes (junctions) in a assigned network.

c) Joining all the said nodes with links (pipes)

d) Google earth map of the study area

Hazen William Formula:

1. Head loss (H_f) = [illegible] (i)
2. Velocity (V) = $0.354CD^{.63}S^{.54}$ (ii)
3. Discharge(Q) = $0.278CD^{2.63}S^{.54}$ (iii)

Where,
D = Diameter (m)
H_f = Head Loss (m)
Q = Discharge (m³/ sec)
V = Velocity of flow (m/s)
S = Hydraulic Gradient (m/m)
L = Length of the pipe (m)
C = Hazen William Coefficient (as per material)

- WDN is be solved using the **Hazen William's formulae** as it is the best-suited method for solving the gravity based water distribution networks.
- Hydraulic software **EPANET** is used for determining discharge, velocity, headloss, etc.

Results

With the help of simulation, discharge, velocity and head loss of the proposed water network were found out along with respective graphs of all parameters vs. time is created which gives a better understanding of the pattern. (**Ref.Fig.5**)

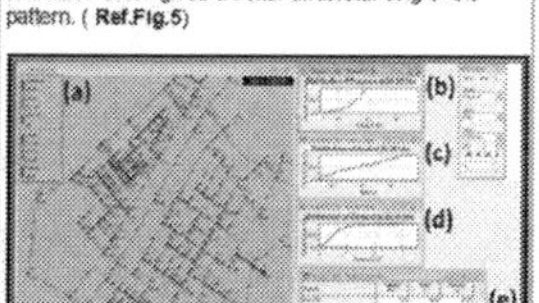

Fig 5: RESEARCH PARAMETERS & GRAPHICAL RESULTS

(a) Simulation shows pressure & flow distribution

(b) Frequency plots for pressure distribution

(c) Frequency plots for head distribution

(d) Frequency plots for demand distribution

(e) Nodal table showing demand ,head and pressure.

Conclusion

- **Times series plot** were made for all nodes from 24 hours range demand pattern which helped in analysis of WDN. Frequency plots for distribution of pressure, head and demand at any interval of time shows that if **DEM** is considered flow through gravity and aspects such as **optimization, maintenance** can be achieved (**Fig.5 a & b**)
- Aplication of GIS suggested a proper smart network which plays a vital role in economy

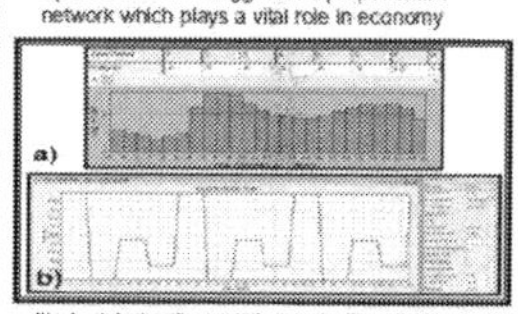

Fig 5: a) 24 hrs Demand Pattern b) Time Series Plot

- For each of the node as well as pipe various graphs relating to pressure & flow in iron pipes (**Fig.6**) and others materials also.

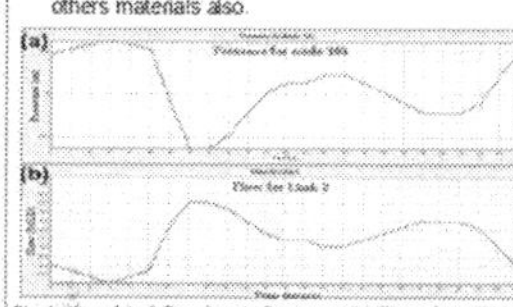

Fig 6. Graphical Results :(a)Pressure (b) Flow- Iron Pipes

References

1. Ji Soo Lee ,Won Hwa Hong,**(2014)**"Development of small district adjustment on public water demand to decrease municipal water supply load". Urban Water ,13(2), Sept.2014, pp 142-155
2. K. Vairavamoorthy a,b, Jimin Yana , Harshal M. Galgale a , Sunil D. Gorantiwar, **(2006)**, "R16 IRA-WDS A GIS-based risk analysis tool for WD system", Elsevier, Environmental Modelling & Software ,Vol 22 7, Jan 2007 , pp 951-965
3. Baiyi Jiang ,Fabei Zhang , Jian Gao and Hongbin Zhao,**(2012)** " Building Water Distribution Network Hydraulic Model by Using WaterGEMS" ASCE, International Conference on Pollution and Treatment Technology,ITCPTT, Vol1 , Oct 2012 , pp 453-461
4. A Baminu, Abubakar, Nasiru, L Sagir**(2013)** "Design of NDA Water Distribution Network Using EPANET" International Journal of Emerging Science and Engineering (IJESE) ISSN, 1(9),2013,pp 2319–6378

- **Website References:**
- https://cfpub.epa.gov (EPA Science Inventory)
- https://bhuvan.nrsc.gov.in/bhuvan_links.php (Bhuvan)
- http://www.lib.utexas.edu/maps/ams/india (Topo sheet)
- http://www.qgis.org (QGIS)
- http://epanet.de/ghydraulics/index.html.en (Plug-in)
- https://plugins.qgis.org/plugins/Qgis2threejs/ (Plug-in)
- https://www.epa.gov/water-research/epanet (Epanet),
- www.waterworld.com (water world)

Aknowledgements

Thanks
Dr.Shrikant Charhate (Guide)
IIRS Faculty and Scientists ,Dehradun
Pillai HOC College of Engineering and Technology , Rasayani
My Parents and well wishers

38th Asian Conference on Remote Sensing
Space Applications: Touching Human Lives
October 23-27, 2017 New Delhi (India)

INDIAN INSTITUTE OF REMOTE SENSING
Indian Space Research Organisation
Department of Space, Government of India

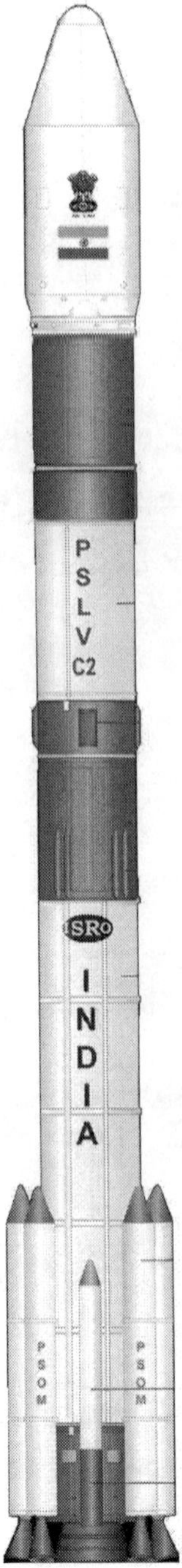

Water Distribution Network Utility Management Using Geospatial Techniques and Hydraulic Simulation – A Remote Sensing and GIS Approach

Project submitted at IIRS, ISRO, Dehradun in partial fulfillment of the requirement of the award of

NNRMS 2017

Submitted by:

Mr. Karthik Nagarajan

Supervised by:

Dr. Praveen K.Thakur

Mr. Kamal Pandey

Indian Institute of Remote Sensing, ISRO,

Department of Space, Government of India

Dehdradun – 248001, Uttarakhand, India

June 2017

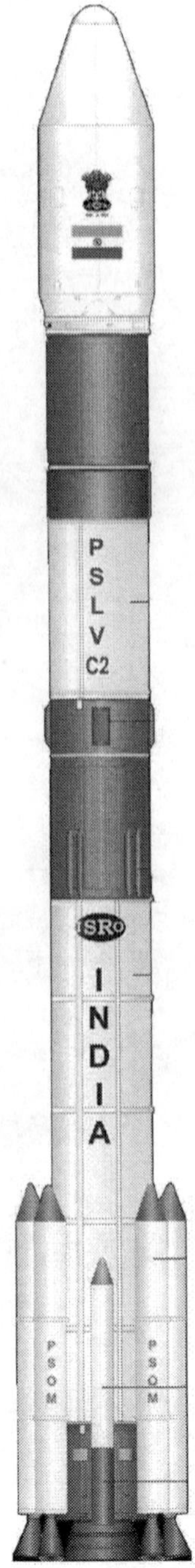

Indian Institute of Remote Sensing, ISRO, Dehdradun

Water Distribution Network Utility Management Using Geospatial Techniques and Hydraulic Simulation – A Remote Sensing and GIS Approach

Project submitted at IIRS, ISRO, Dehradun in partial fulfillment of the requirement of the award of
NNRMS 2017

Submitted by

Mr. Karthik Nagarajan

Supervised by

Dr. Praveen K. Thakur,

Scientist/Engineer SF WRD Dept.

Dr.Kamal Pandey

Scientist / Engineer SD, GeoWeb Dept.

Indian Institute of Remote Sensing, ISRO,
Department of Space, Government of India
Dehdradun – 248001, Uttarakhand, India

June 2017

भारत सरकार
अंतरिक्ष विभाग
भारतीय सुदूर संवेदन संस्थान
4, कालीदास मार्ग, पो. बाक्स सं. 135
देहरादून– 248 001, भारत
दूरभाष : +91-135-2524399
फैक्स : +91-135-2741987, 2748041

Government of India
Department of Space
Indian Institute of Remote Sensing
4, Kalidas Road, P.B. No. 135,
Dehradun - 248 001, India
Telephone : +91-135-2524399
Fax : +91-135-2741987, 2748041

CERTIFICATE

This is to certify that the project work entitled **"Water Distribution Network Utility Management Using GeoSpatial Techniques and Hydraulic Simulation Approach"** submitted by **Mr. Karthik Nagarajan** partial fulfilment of the requirements for the successful completion of **NNRMS 2017**. The project presented here in this report is an original work of the candidate and has been carried out in Water Resources Department, Indian Institute of Remote Sensing, Dehradun, under the guidance of **Dr. Praveen K. Thakur**, Scientist SF, IIRS, Water Resources Department and **Mr. Kamal Pandey**, Scientist SD, Geoweb Services, IT & Distance Learning, Indian Institute of Remote Sensing, Dehradun, India.

22-6-17

(Dr. Praveen K. Thakur)
Scientist/Engineer 'SF'
Water Resources Department

22/6/17

(Mr. Kamal Pandey)
Scientist/Engineer 'SD'
GeoWeb, IT and Distance Learning

(Dr. S. P. Aggarwal)

Scientist /Engineer 'SG' & Head,
Water Resources Department

सत्यमेव जयते

Indian Institute of Remote Sensing
Indian Space Research Organisation
Department of Space, Government of India

इसरो isro

Enrolment No. 10665

Certificate

This is to certify that **Mr. Karthik Nagarajan** from **Pillai HOC College of Engineering and Technology, Rasayani, Raigad, Maharashtra** has been awarded this certificate on having completed the ISRO-sponsored NNRMS Course on **"Remote Sensing and Geographical Information System in Water Resources"** conducted by this Institute during May 01, 2017 to June 23, 2017.

Course Director	**Group Head, PPEG**	**Dean, Academics**	**Director, IIRS**

Date: June 23, 2017
Place: Dehradun, India

PHOTO GALLERY

Indian Institute of Remote Sensing, ISRO

Dehdradun

Indian Institute of Remote Sensing, ISRO, Dehdradun

Indian Institute of Remote Sensing, ISRO, Dehdradun

With my Guides Dr. Praveen K. Thakur, Dr.Kamal Pandey

Indian Institute of Remote Sensing, ISRO, Dehdradun

A

B

C

E

F

A) **DR. A. Senthil Kumar** *(DIRECTOR IIRS)*
B) **Dr.Suresh Kumar,** Head ASD & Scientist SG, And IIRS, Course Director
C) **Dr. Bhaskar Nikam** , Scientist/Engineer- SE
D) **Dr. S. P. Aggarwal,** Head WRD, IIRS
E) **Dr. Vaibhav Garg** , Scientist/Engineer- SE

All Scientists at Water Resources Department , IIRS, ISRO, Dehdradun

With Research Scholars at Water Resources Department, IIRS, ISRO, Dehdradun

In Situ Ground Water Detention Experiment
Ground Elevation Survey

Global Positioning System

Digital Map showing Latitude and Longitude across world

Construction details of Satellite, Sensors, and Payloads

Geo Referencing with the use of TopoSheet and LULC Maps

Various Payloads and Satellite launching devices

Seminar on Application of RS & GIS in Water Resources Management and Civil Engineering

PSLV Model

Launching Pad Mechanism

Centre of Space Science and Technology Education in Asia and the Pacific

OUTREACH FACILITY
NORTH EASTERN SPACE APPLICATIONS CENTRE

ISO 9001:2015 Certified
NORTH EASTERN
SPACE APPLICATIONS
उत्तर पूर्वी अन्तरिक्ष
Government of India
भारत सरकार
Umiam 793103,
उमियम 793

Technology for Sustain
emphasis on NER
Indian Society of Geomatics & Indian Society of Remote

Valedictory Function
Conducted by:
Indian Institute of Remote Sensing, Dehradun
and
U R Rao Satellite Centre, Bengaluru
Indian Space Research Organization
Dept. of Space , Govt. of India
at
CSSTEAP

iirs

@karthik_n_iyerr

IIRS Academia Meet - 2020
MARCH 03, 2020

iirs
in Asia and the Pacific (CSSTEAP)
(Affiliated to the United Nations)
Valedictory Function
Conducted by:
Indian Institute of Remote Sensing, Dehradun
and
U R Rao Satellite Centre, Bengaluru
Indian Space Research Organization
CSSTEAP

हॅटट्रिकचा 'आविष्कार'... मुंबई विद्यापीठाच्या कलिना कॅम्पसमध्ये पार पडलेल्या १४व्या आंतरविद्यापीठ 'आविष्कार' संशोधन स्पर्धेत मुंबई विद्यापीठानं नेत्रदीपक कामगिरी केली. एकूण ९३ गुणांसह मुंबई विद्यापीठानं अंतिम विजेतेपद पटकावलं. यावेळी विद्यापीठानं १५ सुवर्ण आणि ६ रौप्य पदकांची घसघशीत कमाई करत आविष्कार संशोधन स्पर्धेवर सलग तिसऱ्या वर्षी विजयी मोहोर उमटवली. यामध्ये मानव्यविद्या, भाषा आणि ललित कला, वाणिज्य, व्यवस्थापन आणि विधी, मुलभूत शास्त्रं, शेती व पशुसंवर्धन, अभियांत्रिकी आणि तंत्रज्ञान, वैद्यकशास्त्र व औषधशास्त्र या विभागांमध्ये विद्यापीठातर्फे एकूण ४८ संशोधक सहभागी झाले होते. या विजेत्यांचं कौतुक करण्यासाठी कुलगुरू डॉ. सुहास पेडणेकर, प्र-कुलगुरू डॉ. रवींद्र कुलकर्णी आणि विद्यार्थी विकास मंडळाचे संचालक डॉ. सुनिल पाटील उपस्थित होते.

पदकविजेत्या तरुण संशोधकांविषयी...५

मुंबई TIMES

समाजहिताचा आविष्कार

Life is more fun with better hearing

FREE HEARING CHECK UP

IIRS Academia
Meet - 2020
CAPACITY BUILDING IN EMERGING
REMOTE SENSING TECHNOLOGIES:
REQUIREMENTS AND OPPORTUNITIES
MARCH 03, 2020
Universities Press
FUNDAMENTALS OF
REMOTE SENSI
Second Edition
George Joseph

3D Model created and simulated in a Computer Software for visual demonstration

Site Visit at Dapoli

Water Transmission Network

Meeting with CIDCO officers at Raigad Bhavan

WATER TRANSMISSION ANALYSIS FROM SOURCE TO DESTINATION-A GIS AND REMOTE SENSING APPROACH

Made in the USA
Las Vegas, NV
19 September 2026

9ece73e1-8778-43c7-8e88-7f72ee887fafR01